Surrounded by Science: Learning Science in Informal Environments

Marilyn Fenichel and Heidi A. Schweingruber

Board on Science Education

Center for Education
Division of Behavioral and Social Sciences and Education

NATIONAL RESEARCH COUNCIL
OF THE NATIONAL ACADEMIES

THE NATIONAL ACADEMIES PRESS
Washington, D.C.
www.nap.edu

THE NATIONAL ACADEMIES PRESS 500 Fifth Street, N.W. Washington, DC 20001

NOTICE: The project that is the subject of this report was approved by the Governing Board of the National Research Council, whose members are drawn from the councils of the National Academy of Sciences, the National Academy of Engineering, and the Institute of Medicine. The members of the committee responsible for the report were chosen for their special competences and with regard for appropriate balance.

This study was supported by Grant No. ESI-0348841 between the National Academy of Sciences and the National Science Foundation with support from the National Institute of Child Health and Human Development and also the Merck Institute of Science Education. Any opinions, findings, conclusions, or recommendations expressed in this publication are those of the author(s) and do not necessarily reflect the views of the organizations or agencies that provided support for the project.

Library of Congress Cataloging-in-Publication Data
or
International Standard Book Number 0-309-0XXXX-X
Library of Congress Catalog Card Number 97-XXXXX

Additional copies of this report are available from the National Academies Press, 500 Fifth Street, N.W., Lockbox 285, Washington, DC 20055; (800) 624-6242 or (202) 334-3313 (in the Washington metropolitan area); Internet, http://www.nap.edu.

Printed in the United States of America

Suggested citation: Fenichel, M., and Schweingruber, H.A. (2010). *Surrounded by Science: Learning Science in Informal Environments.* Board on Science Education, Center for Education, Division of Behavioral and Social Sciences and Education. Washington, DC: The National Academies Press.

THE NATIONAL ACADEMIES

Advisers to the Nation on Science, Engineering, and Medicine

The **National Academy of Sciences** is a private, nonprofit, self-perpetuating society of distinguished scholars engaged in scientific and engineering research, dedicated to the furtherance of science and technology and to their use for the general welfare. Upon the authority of the charter granted to it by the Congress in 1863, the Academy has a mandate that requires it to advise the federal government on scientific and technical matters. Dr. Ralph J. Cicerone is president of the National Academy of Sciences.

The **National Academy of Engineering** was established in 1964, under the charter of the National Academy of Sciences, as a parallel organization of outstanding engineers. It is autonomous in its administration and in the selection of its members, sharing with the National Academy of Sciences the responsibility for advising the federal government. The National Academy of Engineering also sponsors engineering programs aimed at meeting national needs, encourages education and research, and recognizes the superior achievements of engineers. Dr. Charles M. Vest is president of the National Academy of Engineering.

The **Institute of Medicine** was established in 1970 by the National Academy of Sciences to secure the services of eminent members of appropriate professions in the examination of policy matters pertaining to the health of the public. The Institute acts under the responsibility given to the National Academy of Sciences by its congressional charter to be an adviser to the federal government and, upon its own initiative, to identify issues of medical care, research, and education. Dr. Harvey V. Fineberg is president of the Institute of Medicine.

The **National Research Council** was organized by the National Academy of Sciences in 1916 to associate the broad community of science and technology with the Academy's purposes of furthering knowledge and advising the federal government. Functioning in accordance with general policies determined by the Academy, the Council has become the principal operating agency of both the National Academy of Sciences and the National Academy of Engineering in providing services to the government, the public, and the scientific and engineering communities. The Council is administered jointly by both Academies and the Institute of Medicine. Dr. Ralph J. Cicerone and Dr. Charles M. Vest are chair and vice chair, respectively, of the National Research Council.

www.national-academies.org

OVERSIGHT GROUP ON LEARNING SCIENCE IN INFORMAL ENVIRONMENTS PRACTITIONER VOLUME

SUE ALLEN, Visitor Research and Evaluation, Exploratorium, San Francisco
MYLES GORDON, Myles Gordon and Associates, Hastings on Hudson, New York
LESLIE HERRENKOHL, College of Education, Cognitive Studies in Education and Human Development and Cognition, University of Washington, Seattle
GIL NOAM, Program in Afterschool Education and Research, McLean Hospital and Harvard Medical School
NATALIE RUSK, Media Lab, Massachusetts Institute of Technology
BONNIE SACHATELLO-SAWYER, Hopa Mountain, Inc., Bozeman, Montana
DENNIS SCHATZ, Pacific Science Center, Seattle

C. JEAN MOON, *Director (until October 2007)*
HEIDI A. SCHWEINGRUBER, *Deputy Director*
ANDREW W. SHOUSE, *Senior Program Officer*
MICHAEL A. FEDER, *Senior Program Officer*
THOMAS KELLER, *Senior Program Officer*
KELLY DUNCAN, *Senior Program Assistant*

Acknowledgments

This book would not have been possible without the sponsorship of the Institute for Museum and Library Studies, the National Science Foundation, and the Burroughs-Wellcome Fund.

We want to thank the following practitioners for their insightful feedback on an early draft of the report: Rick Bonney, Program Development and Evaluation, Cornell Lab of Ornithology; Chris Gentile, Oatland Island Education; Julie Johnson, Science Museum of Minnesota; Martiza MacDonald, American Museum of Natural History; Jennifer Martin, Children's Discovery Museum of San Jose, California; Ellen McCallie, Center for Advancement of Informal Science Education, Association of Science-Technology Centers; Dale McCreedy, Gender and Family Learning Programs, Franklin Institute Science Museum; Kathy McLean, Independent Exhibitions; Thea Sahr, Educational Outreach, WGBH; Mary Ann Steiner, Center for Learning in Out of School Environments, University of Pittsburgh; and Gretchen Walker, Community and Visitor Programs, Lawrence Hall of Science, University of California, Berkeley. They provided constructive comments and suggestions that were essential to the project.

This report has been reviewed in draft form by individuals chosen for their diverse perspectives and technical expertise, in accordance with procedures approved by the Report Review Committee of the National Research Council (NRC). The purpose of this independent review is to provide candid and critical comments that will assist the institution in making its published report as sound as possible and to ensure that the report meets institutional standards for objectivity, evidence, and responsiveness to the charge. The review comments and draft manuscript remain confidential to protect the integrity of the deliberative process.

We wish to thank the following individuals for their review of this report: Bronwyn Bevan, Informal Learning and Schools, The Exploratorium, San Francisco, CA; Rick Bonney, Program Development and Evaluation, Cornell Lab of Ornithology; Lynn D. Dierking, Free-Choice Learning and Department of Science and Mathematics Education, Oregon State University; Cecelia Garibay, President's Office, Garibay Group, Chicago, IL; Chris Gentile, Office of Director, Oatland Island Wildlife Center, Savannah, GA; Thea Sahr, Special Initiatives, WGBH-TV, Boston, MA; and Gretchen Walker, Community and Visitor Programs, Lawrence Hall of Science, University of California, Berkeley

Although the reviewers listed above provided many constructive comments and suggestions, they were not asked to endorse the conclusions and recommendations nor did they see the final draft of the report before its release. The review of this report was overseen by Cary I. Sneider, College of Liberal Arts and Sciences, Portland State University. Appointed by the NRC, he was responsible for making certain that an independent examination of this report was carried out in accordance with institutional procedures and that all review comments were carefully considered. Responsibility for the final content of this report rests entirely with the authoring committee and the institution.

Contents

Preface

As children, many of us remember going on a family outing to a zoo, an aquarium, a planetarium, or a natural history museum. Although sometimes we may have said it was boring, eventually there was something that caught our eye. Perhaps it was a chimpanzee staring back at us in a strangely familiar way or a shark taking a solitary swim in a custom-made tank. It could have been a moon rock brought back to Earth from one of the first manned space flights.

When, at the end of the outing, parents asked, "Did you have fun?" in spite of ourselves we usually had to say yes. But then they wanted to know something else: "What did you learn?" That question was far harder to answer.

Indeed, those working in science museums and other informal learning environments, including film and broadcast media; botanical gardens and nature centers; and youth, community, and out-of-school-time programs, increasingly are being called on to answer this question. Although people have participated in these activities for at least 200 years, only in the last few decades have practitioners and evaluators in the informal science community begun to systematically study what people learn, how they learn, and whether experiences in informal environments reinforce people's identity as science learners. This work, still in its early stages, has proven to be quite challenging, for several reasons.

For one thing, ideas about learning have become increasingly sophisticated. It turns out that learning is far more than simply accumulating content knowledge. It is also a social process, informed and enhanced by collaboration and discussion with other learners. In addition, "science learning" has its own particular characteristics. It encompasses the building of conceptual knowledge as well as mastering skills, such as observing, making predictions, designing experiments, and drawing conclusions based on data. What's more, science learning also has a cultural component. Science has its own language, tools, and practices. Part of the learning process for nonscientists is becoming familiar with the culture of science and figuring out how it meshes with their own cultural perspectives.

Scientists constantly revise their understanding of how the world works based on new evidence emerges. For example, until recently, everyone considered Pluto to be a planet, but now the best minds in astronomy say otherwise. The Mars Land Rover program has brought back pictures showing that Mars once had pockets of water, confirming what had previously been only a hypothesis. And in the field of biology, there has been a shift in focus, moving from an emphasis on the structure and function of plants and animals to one on molecular and cell biology.

Many compelling current issues are related to scientific knowledge, which provides the background needed to make decisions about problems and to take advantage of opportunities. For example, although science cannot tell people what to do about climate change, it can provide the data necessary to realize that carbon dioxide emitted into the air, often through human activities, is greatly affecting the climate. The way people interpret that information—and whether they accept it—is based on their cultural context, values, and vision for the future. The same holds true for acceptance of a new avenue of study, such as stem cell research. Science presents the opportunity to pursue it, but people's beliefs and values dictate whether they follow through.

One of the goals of informal science environments is to introduce learners to scientific skills and concepts, the culture of science, and the role science plays in decision making. While some of this can be learned in school, informal settings have the advantage in that they can reach people of all ages, with varying levels of interest and knowledge of science. What tools and strategies are needed to help practitioners in informal settings meet these challenges? What knowledge could help inform their practice?

This book strives to answer these questions. One of its key premises is that an understanding of current research about how people learn in general--as well as the specific challenges of learning science—can improve the quality of informal science offerings. For example, exhibits can become more interactive, which research says has the potential to provoke questions and elicit more thoughtful comments and conversations. Strategies used in commercially produced computer games can be put to use in "educational" games to generate excitement about science as well as to build players' knowledge base. And out-of-school-time programs, especially those for nondominant groups, can be designed with an understanding of the participants' culture.

These findings and others brought together in this book come from the study report, *Learning Science in Informal Environments: People, Places, and Pursuits.* This report, written by a committee of 14 experts convened by the National Research Council, includes the perspectives of developmental and cognitive psychologists, science educators, museum researchers and evaluators, social scientists, and professionals in the fields of youth and adult learning. This group reviewed the most relevant peer-reviewed research, commissioned new papers on specialized topics, and held three public fact-finding meetings. Their report distilled what is known from research while also identifying what gaps remain in our knowledge about how to create effective informal science learning environments.

Along the way, the committee realized that its findings would have tremendous value to a wide range of practitioners: educators, museum professionals, policy makers, university faculty, youth leaders, media specialists, publishers, and broadcast journalists are among those who could put these new insights to good use. As a result, this book was created with several purposes in mind: to introduce newcomers to the growing body of research, to enhance the knowledge base of mid-level professionals, and to provide seasoned professionals with a source that gathers the body of research together in an accessible format. For all these audiences, the goal is to present what the committee sees as the best thinking to date on how people learn in informal science environments.

The book is divided into three parts. Part I, Frameworks for Thinking about Science Learning, lays the foundation for much of the research referred to throughout the book. The first chapter describes the range of informal environments for learning science, including everyday environments, designed environments, and programs, and then makes the point that those environments developed by professionals share common commitments. Among others, these encompass a desire to engage participants in multiple ways, to provide opportunities for direct interaction with phenomena, and to build on learners' prior knowledge and interests. Chapter 2 builds on the discussion by focusing specifically on what it means to do and learn science. The chapter opens with a discussion of science as a human endeavor that involves specialized language, tools, and norms. It then introduces the strands of science learning, a framework that describes the range of knowledge, skills, interests, and practices involved in science learning. The strands framework is a tool that can be used to reflect on the broad range of competencies involved in learning science, to articulate learning goals, and to guide evaluation. The strands come up throughout the book in the descriptions of different types of informal environments and the type of learning that has occurred.

Part II, Designing Experiences to Promote Science Learning, focuses on different aspects of the research on learning and how it can be put to work by practitioners, as well as assessment. Chapter 3 discusses specific strategies, such as the use of interactivity, that are effective in building the deeper, more flexible understanding of science that is exemplified by the strands. Chapter 4 highlights the social and cultural aspects of learning, exploring how individual learning is supported through interaction with more knowledgeable others and through the dynamic exchange of ideas. Chapter 5 discusses ways to enhance interest and motivation to learn and how a developed identity as a science learner is both a natural outcome of a highly motivated learner and a reason that people pursue varied informal learning experiences in science. Part II concludes with a chapter that explores the role of assessment in informal settings and the challenges inherent in this endeavor.

Part III, Reaching Across Communities, Time, and Space, emphasizes other variables that affect learning. Chapter 7 presents a detailed discussion of what is meant by "equity" in the context of informal science settings and how these environments can be made more accessible to diverse populations. Chapter 8 discusses how to develop effective learning experiences for learners across the life span--children and youth, senior citizens, and other adults. Chapter 9, the final chapter in the book, looks to the future of informal science learning, with a discussion on how to extend learning experiences across different media and settings. It also looks at the relationship between formal and informal science environments and discusses the value to the learner of creating stronger links between these two communities.

Throughout the book, case studies show how the principles and strategies emerging from research on learning can and are being employed by informal science educators across various settings. They also provide concrete examples to reflect on and critique, with the hope that they will generate new insights that will inform readers' own work. For those who want to pursue the topics presented in each chapter in greater depth, a list of additional readings is included. Also, there is a list of "things to try" that provides suggestions for how to take ideas discussed in the chapter and begin to apply them.

A major goal of the book is to show the many ways that informal environments can support science learning and particularly to provide insight into how science can be made meaningful to people of all ages, backgrounds, and cultures—a value long held dear in the informal science community. Columbia University physicist Brian Greene offers an eloquent explanation of this belief: "Science is a way of life. Science is a perspective. Science is the process that takes us from confusion to understanding in a manner that's precise, predictive and reliable—a transformation, for those lucky enough to experience it, that is empowering and emotional. To be able to think through and grasp explanations—for everything from why the sky is blue to how life formed on earth—not because they are declared dogma but because they reveal patterns confirmed by experiment and observation, is one of the most precious of human experiences."

Through informal science learning, we all can experience this joy as our eyes are opened to the excitement and wonder that is science.

Part I
Frameworks for Thinking About Science Learning

1
Informal Environments for Learning Science

A great deal of science learning, often unacknowledged, takes place outside school in museums, nature centers, after-school programs, amateur science clubs, and even during conversations at the dinner table. Collectively, these kinds of settings are often referred to as informal learning environments. Understanding the science learning that occurs in these environments in all its complexity and then exploring how to more fully capitalize on these settings for learning science are the major issues addressed in this book.

Virtually all people of all ages and backgrounds engage in informal science learning in the course of daily life. In fact, despite the widespread belief that schools are responsible for addressing the scientific knowledge needs of society, the reality is that schools cannot act alone. Society must better understand and draw on informal experiences to improve science education and science learning broadly.

Consider, for example, that by some estimates individuals spend as little as 9 percent of their lives in schools.[1] Furthermore, science in K-12 schools is often marginalized by traditional emphases on mathematics and literacy, hence little science is actually taught during school hours.[2] Finally—although it needn't be and isn't always so—much of science instruction in school focuses narrowly on the "facts" of science and simplistic notions of scientific practice.[3] Yet the growing body of research on science learning makes clear that a basic understanding of science requires far more than the acquisition of a body of science knowledge. Rather, knowing science includes understanding, at a basic level, the nature and processes of science. For these reasons, now more than ever, informal environments can and should play an important role in science education.

VENUES FOR LEARNING SCIENCE

As individuals interact with the natural world and participate in family and community life, they develop knowledge about nature and science, as well as science-relevant interests and skills. Science learning can occur through a number of experiences, including mentorship, reading scientific texts, talking with experts, watching educational television, or participating in science-related clubs. Informal learning experiences are often characterized as being guided by learner interests, voluntary, personal, deeply embedded in a specific context, and open-ended.[4] Successful informal science learning experiences are seen as not only leading to increased knowledge or understanding in science, but also to further inquiry, enjoyment, and a sense that science learning can be personally relevant and rewarding.

In order to make sense of the vast number of informal settings in which science learning might occur, we use three categories developed in the National Research Council report *Learning Science in Informal Environments: People, Places, and Pursuits*.[5] These include everyday informal environments (such as watching TV, reading books, having conversations, pursuing one's hobby, volunteering for an environmental cause), designed environments (such as museums, science centers, aquariums, zoos, environmental centers, visitor centers of local, state, or national parks), and programs (such as 4-H programs, museum science clubs, citizen science activities, after-school activities). All of these environments can be placed on a continuum characterized by the degree of choice given to the learner or group of learners, the extent to which the environments and experiences provided are designed by people other than the learners, and the type and use of assessments.

Everyday Learning

Everyday learning includes a range of experiences that may extend over a lifetime, such as family discussions, walks in the woods, personal hobbies, watching TV, reading books or magazines, surfing the Web, or helping out on the farm. These experiences are very much selected and shaped by the learners themselves and may vary greatly across families, communities, and cultures. People engaging in everyday learning may not be aware that they are learning. Instead, they simply see the activity as part of their daily lives—engaging in a hobby, looking up information on the Internet, enjoying a science documentary on TV, reading a fascinating book about the life of Darwin, playing games (in the backyard, at home, or on the computer) or having a meaningful conversation with friends.

Consequently, learners may not be explicitly asked to demonstrate competence in the same way they are when tested in school. Rather, demonstration of competence or signs of learning are embedded in the activity—for example, parents praising a child who explains how a tree "drinks" water or friends correcting and challenging each other when discussing which foods are the healthiest to eat. In informal settings, individuals may take on or are given more responsibility or more difficult tasks when it is clear that their competence has increased. For example, a child growing up in an agricultural society may start with feeding animals and cleaning stalls and gradually assume responsibility for tending animal wounds and monitoring the animals' well-being. An amateur astronomer may take on increasingly more sophisticated outreach tasks, progressing from aiding at a public star party to delivering a lesson on astronomy to schoolchildren.

Designed Environments

Designed environments include museums, science and environment centers, botanical gardens, zoos, aquariums, visitor centers of all kinds, historic settings, and libraries. In these settings, artifacts, media, signage, and interpretation by staff or volunteers are primarily used to guide the learner's experience. When the environments themselves are structured by staff of the institutions, individual learners and groups of learners determine for themselves how they interact with them. The choice to attend a museum, aquarium, zoo, or other designed environment is made by the learner or by an adult supervising the learner (e.g., a parent or teacher). Once in the setting, learners have significant choice in selecting their own learning experiences by choosing to attend to only those experiences or exhibits—or aspects of them—that align with their interests. Typically, learners' engagement is short-term and sporadic in these environments, and learning can take place individually or in peer, family, or mentor interactions. However, there is increasing interest in extending the impact of these experiences over time through post-visit web experiences, traveling exhibits, and follow-up mail or e-mail contact. These kinds of innovations are discussed in more detail in Chapter 8.

Programs

Programs include after-school programs, summer programs, clubs, science center programs, Elderhostel programs, volunteer groups, citizen science experiences, science cafes, public lecture series, and learning vacations. Often program content includes a formal curriculum that is organized and designed to address the concerns of sponsoring institutions. Although the curriculum and activities are focused primarily on content knowledge or skills, they may also address attitudes and values and may use science to solve applied problems. Often, programs are designed to serve those seen to be in need of support, such as economically disadvantaged children and adults.

As in designed environments, individuals most often participate in programs either by their own choice or the choice of a parent or teacher. They attend programs that align with their interests and needs. Experiences in these environments are typically guided and monitored by a trained facilitator and often include opportunities for collaboration. The time frame of these learning experiences ranges from brief, targeted, short-term experiences to sustained, long-term programs with in-depth engagement. Assessments are often used to determine progress and to allow for adjustments, but they are not typically meant to judge individual attainment or progress against an objective standard or to form the basis for graduation or certification of any kind (although they may affect the participants' reputation or status in the program or their self-perception and self-confidence).

Insights About Learning in These Environments

Although these three types of environments are very different, they all share some basic characteristics that are believed to encourage learning:

- engaging participants in multiple ways, including physically, emotionally, and cognitively;
- encouraging participants' direct interactions with phenomena of the natural world and the designed physical world in ways that are largely directed by the learner;
- providing multifaceted and dynamic portrayals of science; and
- building on the learner's prior knowledge and interest.

These characteristics have emerged from a philosophical stance toward what it means to provide an informal experience, and they also are informed by a growing research base on learning and how best to promote it. This research base, which forms the foundation for this book, represents multiple fields of inquiry that reflect a wide range of interests, questions, and methods. This diversity of approaches to investigating learning outside schools—both how it occurs and how best to support it—makes the evidence base difficult to pull together. At the same time, the research reveals that the opportunities for promoting learning, as well as inherent challenges in doing so, are similar across informal environments.

ILLUSTRATING THE COMMON CHARACTERISTICS OF INFORMAL ENVIRONMENTS

Two examples provide insight into different kinds of informal learning experiences. One is a computer game that can be played at home (an everyday learning experience), and the other is a program for adults. They occur in different settings, with different age groups, different structures, and different time scales. The similarities and differences in the two descriptions highlight the shared characteristics of informal environments for science learning as well as the unique potential for learning that variation in design can provide.

[CASE STUDY 1-1]

WolfQuest: Playing to Learn

Imagine the opportunity to explore a swath of Yellowstone National Park not from our human perspective but as a wolf. From this vantage point, learning how to hunt and get along with other wolves is a matter of life and death, and the purpose of the game is to make the player learn about wolf behavior and the environment as a means to succeed in this educational equivalent of a multiplayer first shooter game.

Players enter the world of *WolfQuest Episode 1: Amethyst Mountain* as wolf avatars to find out firsthand what it is like to use their senses to track elk, pick out a "good" elk (one that is a little weaker than the rest), and then chase and hunt it down. Defending a carcass against grizzly bears and other competitors also is part of a day's work. Players can go it alone or join a pack with their friends—but if they do that, they have to learn how to cooperate with other members of the pack.

Much to the delight of the game's developer, David Schaller, and his partner from the Minnesota Zoo, Grant Spickelmier, players' responses to the game have exceeded their expectations. "There is a following for our game," says Schaller. "In fact, even before the game was launched, a few teenagers saw an announcement about the game on the Zoo's Web site and posted links to our site on *Zoo Tycoon* and *My Little Pony* game forums. About 4,000 people downloaded the game in the first hour after it was launched, and another 250,000 have downloaded it since. These kids have, in fact, built a community."

One of the ways that this community stays vibrant is through an online forum. Through their posts, kids wax eloquent about everything from the game development process to questions about wolves and places to go for more information. "Kids got so excited about the game that they sent in drawings and stories about wolves," remarks Spickelmier.

To ensure that forum participants stay on task, a moderator gently guides the conversation by posting provocative research findings and facilitating productive discussions. For example, participants had many conversations about whether wolves were going to remain on the list of endangered species or be removed. The job of keeping conversations moving in a constructive direction takes between 15 and 20 hours of paid staff time each week. A team of 18 volunteer moderators, drawn from the older members of the *WolfQuest* community, provides support by filling in for the moderator when she is not working.

Based on evaluations by the game developers, there is evidence of learning. After analyzing forum postings, the developers found that the data appear to be pointing to the use of inquiry behaviors. Players made predictions about what hunting and mate-finding strategies might work, tested those predictions, analyzed the results through the use of observation and note-taking skills, and worked with their pack mates to develop new strategies. One player analyzed his maneuvers as follows: "I had to overcome speed, and I had trouble with social behavior. I also had to keep up with hunting, and trying not to die. Survival of the fittest. I tested being dominant over the stranger wolves, and how to save energy for hunt. Once I found a mate, everything got easier."

Schaller and Spickelmier also discovered that the players sought out additional wolf-related experiences as a result of their experience with the game. More than 80 percent of participants looked up information on the Internet; watched a television show or video about wolves; read about wolves in books, magazines, or newspapers; or talked about the game with family and friends. About 70 percent of the players visited a zoo, nature center, or park to actually see wolves and other wildlife. Interestingly, it appears that the more frequently individuals played the game, the more likely they were to engage in one of these follow-up behaviors.

Extrapolating from these findings, it appears that game-playing has potential as a tool that can be used to build knowledge and inquiry behaviors and even lead to additional activities related to wolves and nature. These learning gains, Spickelmier notes, happen as part of the game. As intended, the kids don't even realize that they are doing science. They're just trying out different ways to make their wolves successful in their environment.

> "In the world of games in which these kids have grown up," says Spickelmier, "they expect to have some control over their learning. Maybe that's why they like games so much. For them, the ability to manipulate their environment is the way education is done."

(Adapted from interviews with Grant Spickelmier and David Schaller and from the following evaluation report: Schaller, D. et al., Learning in the Wild: What WolfQuest Taught Developers and Game Players, in J. Trant and D. Bearman (eds.). Museums and the Web 2009: Proceedings, Toronto: Archives & Museum Informatics. Published March 31, 2009. Consulted March 31, 2009. http://www.archimuse.com/mw2009/papers/schaller/schaller.html.

[END OF CASE STUDY 1-1]

This example of everyday learning through media illustrates many of the characteristics of informal learning environments. Participation in the game is entirely voluntary, and the amount of time players devote to it is based on individual choice. The game itself is carefully designed so that, in order to master it, players need to learn about wolves and their habitat. Demonstration of competence is an inherent part of the game, since a player will not succeed unless he or she learns about wolves and is able to use that knowledge to inform his or her strategy and choices in the game.

The second program, called Science Café, was developed by Boston's public television station WGBH. Unlike WolfQuest, which is for children and teens, Science Café is an evening-long event designed for adults. While Science Cafes have been adapted from Europe and occur nationwide, the particular example reflects the model developed by WGBH, and it takes place in a pub near Boston. The WGBH model distinguishes itself by introducing the topic of discussion with a brief video from WGBH's extensive science documentary material.

[CASE STUDY 1-2]

Science in Unexpected Places: Learning at a Science Café

On a cold November evening in Somerville, Massachusetts, some people ventured out to the neighborhood bar, the Thirsty Scholar Pub, a place that attracts mostly a combination of local working-class people and young professionals from the moderately priced apartments in the area.

Its low lighting, comfortable tables and chairs, and televisions make it a perfect spot for local sports fans. Some of the adults in the audience had tickets for this night's special event, and others just happened to be there.

On this particular night, Ben Wiehe, WGBH outreach project director, had booked the pub for a Science Café, a program designed to bring science to people of all backgrounds. The subject for this evening's Science Café was global warming, and the topic was going to be introduced through a video clip from a NOVA scienceNOW program. The video focused on the cause of a mass extinction that took place 250 million years ago, which ended what is known as the Permian era. Following the video, Charles Marshall, a Harvard professor and curator at the Museum of Comparative Zoology, was going to speak about the topic, relate it to the global warming being experienced today, and then facilitate a question-and-answer session.

After deciding to have a Science Café, the first step in the NOVA scienceNOW Science Café model is to select the target audience for the event and to choose a venue where that audience is comfortable meeting. The next step is to pick a topic, which involves finding a video clip and scientist for the event. Those tasks both fall to Wiehe. "The video clip is very important because it is designed to galvanize the audience," he explains. "The length is key. I try to go with something that is within four minutes."

The choice of the scientist is equally important. As the host of the Science Café, Wiehe takes an active role in facilitating discussion during each event. Nonetheless, it is still important for him to find a scientist with a good sense of humor who can talk about a science topic in a clear, understandable way. "We want someone who knows how to promote conversation in the group. The point of the event is for the participants' voices to be heard," Wiehe explains.

When Wiehe arrived at the Thirsty Scholar Pub an hour early to set up, he noticed a group of ex-Marines sitting near the table he was planning to use for the event. Discovering that they didn't know about the activity planned, he told them about it and suggested that they join in on the conversation. The men agreed to participate. Wiehe had hoped they would stay. One of the purposes of Science Cafés is to reach audiences who would not normally seek out an evening of science. So for this event, Wiehe had reached out to outdoor groups, such as the Sierra Club, which he thought would be interested because of the event's focus on global warming and climate change.

As Wiehe finished setting up, he noticed that the pub was getting crowded. Soon all 80 seats were occupied. He and other WGBH staff members estimated that about 20 percent of the crowd was Thirsty Scholar regulars who had not come for the event. The remaining 80 percent was probably a mix of students and people who had come for the Science Café after hearing about it through fliers, e-mail alerts, friends, or community newspaper announcements.

The evening started with a brief introduction, much of which was drowned out by private conversations and the televisions; then Wiehe ran the video clip. A light and humorous treatment of a serious subject, the clip caught the crowd's attention. It served to introduce the subject to those unfamiliar with it and reinforce knowledge for those who knew something about the topic.

Dr. Marshall followed the video. He had prepared a seven-minute presentation designed to create a link between mass extinction, the subject of the clip, and global climate change. He concluded his brief talk with concerns about dangers to come. As an aside, he mentioned that bovine flatulence contributed to greenhouse gases, a remark that provoked a chuckle in the room. Then he opened the floor for questions and comments. He had a clipboard of notes on hand, thinking he would need to refer to them during the open discussion.

The ex-Marines raised their hands, beginning the conversation by questioning some of Marshall's claims. "How do we know that humans are causing the problem?" they asked. "Are there any beneficial aspects to global warming?" They also challenged what Dr. Marshall had described as people's collective responsibility to protect the planet for future generations. "So what if humans go extinct?" they mused. "Extinctions have happened before. Maybe it's our fate."

Then the audience turned to the issue of bovine flatulence, which Marshall had briefly touched on. "How does bovine flatulence contribute to greenhouse gases?" someone wanted to know. "What if we changed the diet of the cows?" another participant suggested. "If we all became vegetarians, would that help?"

It's a good thing that Marshall has a wry sense of humor—and can think on his feet. For the next 10 minutes, he and the group discussed different ways to deal with this problem. They considered the possibility of feeding cattle different kinds of grains or cutting back on people's weekly meat consumption. Early in the discussion, Marshall cast his notes aside. He hadn't thought that the conversation would go in this direction, so his notes were of little use. He had to draw on his knowledge of this topic to do his part to keep the discussion going.

As it turned out, the event was a learning experience for Marshall, too. In NOVA scienceNOW's national surveys, 38 percent of participating scientists report that their involvement in the program changed the way they present their work to the public.

After 25 minutes of conversation, Wiehe noticed that some people were starting to lose attention. So he ended the group discussion and reminded everyone to enter a prize drawing by completing the evaluation forms at their tables. The noise in the room increased as everyone started talking excitedly with those nearest them. Marshall was immediately surrounded by patrons who had more questions.

Realizing that Marshall was "in demand," Wiehe asked him to circulate around the room. This approach gave everyone the chance to have a face-to-face conversation about whatever interested them most. For some, this meant a technical discussion of the topic. Others simply wanted a chance to meet Dr. Marshall personally. "I'm going to tell my friends I had a beer with a paleontologist," exclaimed one Thirsty Scholar patron. "This event reminded me of how much I love science."

This participant is not alone in his enthusiasm. In national surveys, more than 70 percent of those attending a Science Café report staying more up-to-date with current science as a result of the experience. The evidence indicates that the interest ignited through the event was sustained and incorporated into participants' daily lives.

Throughout the rest of evening, patrons of the pub continued to talk about global warming. The pub's owner, delighted with the outcome of the evening, was eager to be involved in the next event. Charles Marshall also expressed his enthusiasm for the evening and his desire to participate in future Science Cafés.

Perhaps the most telling sign of the evening's success lay in the hands of the ex-Marines: tickets to a concert they had chosen not to attend. They opted instead for an evening of stimulating discussion about science.

(Adapted from "A Burger, a Beer . . . and a Side of Science," Nancy Linde, WGBH Boston)
[END CASE STUDY 1-2]

REFLECTING ON THE CASES

WolfQuest and the Science Café represent two very different informal science learning environments. One is for children and teens, and the other is for adults. One is a computer game that is played at home for as long as the learner is engaged, and the other is an organized one-shot event. The goals of WolfQuest were also very different from those of the Science Café. The developers of WolfQuest were experimenting with the learning opportunities available through gaming; Wiehe and his colleagues were trying to provide an enjoyable evening of conversation about science, with the hope of whetting the participants' appetites for more.

Despite the significant differences between the settings, WolfQuest and the Science Café share an important element that characterizes much of everyday learning in science: learning can be a by-product of entertaining engagement which can be designed to create further interest and a desire to learn more about the issues. The players develop knowledge and skills as a means to succeed in a game, and their success is synonymous with learning at least some science but also with developing positive attitudes toward the topic itself, as exemplified by their growing interest in wolves. The patrons of the Science Café experience the dialogic nature of science and are exposed to a researcher who personalizes science and provides authenticity. In both cases science learning is adapted to the environment: gamers play and pub patrons talk and discuss.

Furthermore, in both the computer game and the Science Café, program directors built on the learners' prior knowledge and interests. Schaller and Spickelmier did so by using the features of gaming that kids enjoy and embedding science content into that framework. Once the players were hooked on the game, they began learning the science content. Wiehe used a video clip to hook the audience and prepare them for Marshall's talk. To further this engagement, both programs connected with the participants in multiple ways. In the case of the Science Café, this was by engaging them in a discussion of a topic that was intellectually stimulating and emotionally provocative. By playing the computer game, the participants were involved physically, by manipulating the computer mouse to make decisions about their wolf avatars; emotionally, by taking on the persona of a wolf; and cognitively, by learning what they needed to know to ensure that their wolves survived. Also, participants were allowed and encouraged to follow their own interests. At the Science Café, Marshall allowed people's interests to direct the conversation, even if it was a topic with which he was less familiar.

These similarities are not a coincidence. They reflect the designers' commitment to providing informal experiences for learning and their knowledge of how to support learning. This knowledge is informed by a growing body of research exploring how people learn across settings and how individuals would like to learn or experience the world in their free time.

A FRAMEWORK FOR THINKING ABOUT LEARNING

Over a century ago, scientists began studying thinking and learning in a more systematic way, taking steps toward what are now called the cognitive and learning sciences. Beginning in the 1960s, advances in fields as diverse as linguistics, computer science, neuroscience, and motivation offered provocative new perspectives on human development and powerful new technologies for observing behavior and brain functions. As a result, over the past 40 years there has been an outpouring of scientific research on the mind and the brain—a "cognitive revolution," as some have termed it.[6] At the same time, applied research and evaluation in informal science learning have exploded and provided the informal science learning profession with many of today's fundamental principles and frameworks, many of them informed by the results of this cognitive revolution.

This huge and growing body of research on learning provides important insights for designing informal environment for learning science, including guidance about how to understand the varied types of learning that occur in informal science environments; how to actively support this learning through designed experiences that directly tap into natural learning processes; how to assess learning in these settings appropriately; and how to improve on existing informal science environments, including long-term programs, one-shot events, and exhibits. In broad brushstrokes the research on learning to date has revealed the importance of understanding both how individual minds work during the learning process and how the social and cultural context surrounding an individual shapes and supports that learning.

Research on individual cognition and learning, attitudinal development, and motivation has provided insight into the development of knowledge, skills, interests, affective responses, and identity. Some of the relevant principles of individual cognition and learning are articulated in the National Research Council report *How People Learn.*[7] These principles include the influence of prior knowledge, how experts differ from novices (those with deep knowledge and understanding of a specific topic versus those with a less developed or naïve conceptual understanding of a topic), and the importance of metacognition, or the ability to monitor and reflect on one's own thinking. These ideas can be used to inform the design of informal science experiences. For example, many museums deliberately juxtapose visitors' prior knowledge with "scientific" ideas that can explain the natural phenomena they are engaging with in an exhibit or activity. This approach to design has been shown to help learners question their own knowledge and more deeply reconstruct that knowledge in a way that comes to resemble that of the actual scientific discipline.

The sociocultural perspective explores how individuals develop and learn through their involvement in cultural practices, which encompass the language, tools, and knowledge of a specific community or social group. This area of research grew out of concern that an emphasis solely on learning processes within individual mind's overlooked the crucial role of social interaction, language, and tools in learning. The findings of this research show how social verbal and nonverbal interaction plays a critical role in supporting learning. Importantly, as people develop the culturally valued skills, knowledge, and identities of a specific community, they also bring their own prior experiences and knowledge to their cultural groups. In this way, culture is a dynamic process, shaped and modified by the perspectives of its members. According to this approach, scientists, too, are part of their own cultural group, in which people share common commitments to questions, research perspectives, ideas of what constitutes a viable scientific stance, and how individuals develop effective arguments.

Tools and artifacts are particularly important aspects of the cultural context for learning in science. Scientists use many specialized tools to measure and represent natural phenomena. In addition, tools and artifacts typically represent the backbone of many learning experiences in science. In a museum, for example, visitors make sense of exhibits through forms of talk and physical activities that are fundamentally shaped by the nature of the materials and technological objects they encounter.

Media also represent a rich layer of learning artifacts. Interactive media, multiplayer video games, and television all provide a specific infrastructure for learning. Information has become broadly available through online resources and communities. In fact, many people routinely develop and share media objects that involve sophisticated learning and social interaction. Research and evaluation over the past 10 years have shown the effectiveness of media, but also highlight their limitations. Recognizing affordances and limitations, media and brick-and-mortar experiences are becoming increasingly intertwined—for example, a documentary on the history of the telescope is complemented by a similar full-dome planetarium show, an interactive website that features activities for backyard astronomy exploration and a strategy to link the airing and local release of the shows with outreach activities by amateur astronomy clubs. These kinds of linkages and collaborations are discussed further in Chapter 8.

The chapters of this book draw on this rich body of research to elaborate on how informal environments can best support science learning. Many of the basic principles of learning operate in similar ways across settings. However, different settings and types of experiences have different affordances for learning. For example, a long-term program is likely to support different aspects of learning than a one-shot experience. Similarly, a highly structured exhibit may be more suited for particular kinds of learning outcomes than a purely exploratory one. Such differences mean that practitioners in informal science education need to think carefully about what can reasonably accomplished in their own settings through the experiences they provide.

* * * *

It is clear that a great deal of science learning—often unacknowledged—takes place outside school in informal environments. These environments include the home, while playing on the computer or watching television; in designed spaces, such as science museums; and in the context of out-of-school-time programs or adult-oriented lectures or movies. Although these activities vary considerably and occur in different settings with different age groups, different structures, and different time scales, they all share four common commitments:

- To engage participants in multiple ways, including physically, emotionally, and cognitively;
- To encourage participants' direct interactions with phenomena of the natural and designed world largely in learner-directed ways;
- To provide multifaceted and dynamic portrayals of science; and
- To build on learners' prior knowledge and interests.

These commitments are consistent with findings from research on learning that reveal the importance of understanding both how individual minds work during the learning process and how an individual's social and cultural context shapes and supports that learning. We expand on both aspects of learning in Part II and explore the implications for learning across the range of informal settings. We begin in the next chapter by elaborating on science as a human endeavor and the implications for what it means to learn science.

For Further Reading

Anderson, D. (2003). Visitors' long-term memories of World Expositions. *Curator, 46*(4), 400-420.

Anderson, D., Storksdieck, M., and Spock, M. (2007). The long-term impacts of museum experiences. In J. Falk, L. Dierking, and S. Foutz (Eds.) *In principle, in practice: New perspectives on museums as learning institutions* (pp. 197-215). Walnut Creek, CA: AltaMira Press.

Anderson, D., and Shimizu, H. (2007). Factors shaping vividness of memory episodes: Visitors' long-term memories of the 1970 Japan World Exposition. *Memory, 15*(2), 177-191.

National Research Council (2009). Introduction. Chapter 1 in Committee on Learning Science in Informal Environments, *Learning Science in Informal Environments: People, Places, and Pursuits.* P. Bell, B. Lewenstein, A.W. Shouse, and M.A. Feder (Eds.). Center for Education, Division of Behavioral Sciences and Social Science and Education. Washington, DC: The National Academies Press.

National Research Council (1999). Executive Summary in *How People Learn* (pp. xi-xvii). Committee on Developments in the Science of Learning. Division of Behavioral and Social Sciences and Education. Washington, DC: The National Academies Press.

National Research Council (2008). Four strands of science learning. Chapter 2 in Committee on Science Learning, Kindergarten Through Eighth Grade, *Ready, Set, Science!* S. Michaels, A.W. Shouse, and H.A. Schweingruber (Eds.). Center for Education, Division of Behavioral and Social Sciences and Education. Washington, DC: The National Academies Press.

National Research Council (2007). Goals for science education. Chapter 2 in Committee on Science Learning, Kindergarten Through Eighth Grade, *Taking Science to School.* R.A. Duschl, H.A., Schweingruber, and A.W. Shouse (Eds.). Center for Education, Division of Behavioral and Social Sciences and Education. Washington, DC: The National Academies Press.

Ray, A.G. (2005). *The lyceum and public culture in the nineteenth century United States.* East Lansing: Michigan State University Press.

Yager, R.E., and Falk, J. (Eds.). (2008). *Exemplary science in informal education settings: Standards-based success stories.* Arlington, VA: NSTA Press.

Web Resources

Center for the Advancement of Informal Science Education (CAISE): http://caise.insci.org/
Science Cafés: http://www.sciencecafes.org/

2
Science and Science Learning

A first step in understanding how to promote science learning in informal environments is to develop a full picture of what it means to do and learn science. Over the past few decades, historians, philosophers of science, and sociologists have taken a much closer look at what scientists actually do and have found that the reality differs from common stereotypes. In the conventional view, the lone scientist, usually male and usually white, toils in isolation to understand some aspect of the natural world through a series of controlled experiments. He is removed from the real world, operates in a cerebral world of the mind, and experiences breakthroughs that reveal some "truth" about how the world works.

Studies of what scientists actually do belie these stereotypes. Scientists actually approach problems in many different ways with many different preconceptions. There is no single "scientific method" universally employed by all. Instead, scientists use a wide array of methods to investigate and describe phenomena, and develop hypotheses, models, and formal and informal theories. Nonetheless, they share a common commitment to gathering and using empirical evidence derived from examination of the natural world.

SCIENCE AS A SOCIAL AND CULTURAL ENTERPRISE

Studies also show that science is fundamentally a social enterprise. Science is often conducted by groups or even widespread networks of scientists, and an increasing number of women and minorities are scientists. Scientists communicate frequently with their colleagues, both formally and informally and most active researchers are involved in multiple scientific associations or societies, and multiple collaboration or work groups. They exchange e-mails, engage in discussions at conferences, and present and respond to ideas through publications in journals and books. Scientists also make use of a wide variety of cultural tools, including technological devices, mathematical representations, and methods of communication. These tools not only determine what scientists see, but also shape the kinds of observations they make.

In fact, the scientific community has its own core values, habits of mind, knowledge, language, and other tools. These values include common commitments to questions, research perspectives, and ideas about what a viable scientific stance involves. Making progress in science depends on scientists being open to revising their ideas if called for by the evidence. The complex exchange of information and ideas and eventual evolution in thinking occurs in a community in which scientists also have developed a shared language. This language is added to or modified by scientists from specific disciplines as they work toward their own shared goals.

For example, scientists from different disciplines have developed their own vocabularies, often by giving common words new meanings or by inventing words to describe a new scientific idea or discovery. Biologists, for example, talk about *cells* and *DNA* and *genetics*, and physicists have developed new meanings for such familiar words as *energy*, *force,* and *work.* Scientists in each discipline also depend on specialized tools to carry out their work. Biologists may use tools like optical or electron microscopes to collect information, and astronomers may rely on different kinds of telescopes. Despite these differences, all share the larger goal of accumulating evidence to explore or test their ideas.

Some scholars refer to this collective set of norms, practices, language, and tools as the culture of science. This includes specialized practices for exploring questions through evidence, such as the use of statistical tests, mathematical modeling, and instrumentation, and social practices, such as peer review, publication, and debate. In order to "do" science, people must learn these norms and practices.

There is also another sense in which science is cultural or even political-- science reflects the cultural values of those who engage in it. For example, the choices about what is worthy of attention, different perspectives on how to approach certain problems, and so on are shaped by the cultural values scientists bring with them and sometimes the political and economic environments in which scientific endeavors are funded and sustained. From this latter perspective, as is the case with any cultural endeavor, differences in norms and practices within and across fields reflect not only the varying subject matters of interest, but also the identities and values of the participants. The recognition that science is a cultural enterprise implies that there is no cultureless or neutral perspective on science, nor on learning science—any more than a photograph or a painting can be without perspective. Recognition of both aspects of culture in science is critical for promoting science learning.

Learning to communicate in and with a culture of science is a much broader undertaking than mastering a body of discrete conceptual or procedural knowledge. One observer, for example, describes the process of science education as one in which learners must engage in "border crossings" from their own everyday world culture into the subculture of science.[1] The subculture of science is in part distinct from other cultural activities and in part a reflection of the cultural backgrounds of scientists themselves. By developing and supporting experiences that engage learners in a broad range of science practices, educators can increase the ways in which diverse learners can identify with and make meaning from their informal science learning experiences.

To illustrate how nonscientists can learn to participate in science, we consider the case of Project FeederWatch. This project was specifically designed to help birdwatchers make more scientific and credible observations of birds that appear in their backyards. By interacting with scientists and using the tools of science, birders fine-tuned their observation skills, became more comfortable with the culture of science, and, in some instances, were able to make contributions to the field.

[CASE STUDY 2-1]

Science in Your Backyard: Participating in the Practices of Science

Every November, thousands of avid birdwatchers join the community of fellow birders who participate in Project FeederWatch, one of several citizen science projects operated by the Cornell Lab of Ornithology in Ithaca, New York. The goal of the project is to enhance scientific research by providing a cadre of "citizen scientists" with the opportunity to contribute to science while pursuing their own interests. Participation begins by downloading (or receiving in the mail) the FeederWatch Research Kit, which describes the project goals and rationale, instructions for setting up an observation area, procedures for collecting data, and a subscription to the project newsletter, which includes detailed results of FeederWatch data.

The success of the project depends on the quality of the data submitted from participant observations. To make the data collection process easier, online data collection forms are tailored for each region depending on the types of birds known to be in that area. If the submitted data match expectations, they are automatically added to the database. If unexpected information is reported, such as a bird species outside its usual habitat or expected range, the entry is flagged for review by a project staff member, who looks over the data to see if more information is needed. An e-mail conversation may ensue.

"When those situations occur, we find that the participants are often correct," says Rick Bonney, director of program development and evaluation at the Cornell Lab of Ornithology. "They have observed something we did not know was there, which adds to our overall knowledge base." All participants can access the database and work with the data to answer their own questions as they arise.

This project has been in place for more than 20 years, making the lab staff among the first to give the public an opportunity to collect data and be part of authentic scientific research. The thinking behind the project was that giving "regular" people the chance to engage directly with phenomena and learn how to conduct investigations would help them become comfortable with the tools and practices of science.

Since its inception, thousands of people have participated in this and similar programs. Over the years, staff at the Lab of Ornithology have worked to perfect these programs by conducting regular participant surveys, which are used to develop a profile of the participants and determine which aspects of the program are most popular and how best to ensure that participants are able to make valuable scientific contributions and themselves are well served.

The surveys reveal that typical participants tended to be college-educated white women over age 50 who, despite having watched or fed birds for years, still see themselves as intermediate birders. The vast majority of the participants made use of the website features, such as Rare Bird Reports, the Map Room, the Top 25 list of birds, the Personal County Summaries, or the State/Province Summaries. More than half of participants used such scientific tools as creating trend graphs for specific bird species.

When participants were asked if they have learned about birds from this project, the results were encouraging. About 50 percent said that they learned there was a greater diversity of species than they had known about before; 64 percent said that they had learned to identify more species; 74 percent said that they observed interesting behaviors; and 70 percent said they learned how birds change throughout the seasons. Only 6 percent of the participants said that they didn't learn anything as a result of their involvement in the project.

Comments also show that the project added value to an existing hobby by providing tools that allowed participants to deepen their experience:

> *A participant from North Carolina remarked, "I loved feeding and watching the birds before, but now it is so much more interesting and useful."*

> *A birdwatcher from New Mexico described how the project improved basic birdwatching skills, "After participating in Project FeederWatch for several seasons, my bird identification skills have improved immensely. This winter, I found myself identifying birds by their behavior: how they fly into the feeding site, where they land, if they sit or take right off again, and which feeder they choose."*

Challenging Enthusiastic Birders

Because so many participants return to the program year after year, lab staff have developed additional research projects to give them a chance to engage in deeper inquiry. One project, called the "Seed Preference Test," was designed to find out which of three kinds of seeds ground-feeding birds liked best—sunflower, millet, or milo. The hypothesis developed by the lab staff was that sunflower was the preferred seed, but participants from the Southwest discovered otherwise. The birds in their region loved milo, also referred to as sorghum. Staff were intrigued by this surprising observation and wanted to find out if milo had been getting a bad rap. So they extended the experiment for one additional year.

The research project resulted in a small media buzz. *Good Morning America* featured it on their program, boosting enrollment to more than 17,000 participants. About 5,000 people completed the observations, documenting half a million bird visits and showing seed preferences for more than 30 species. The findings confirmed the reports from the Southwest about seed preferences for birds in their area, proving that the lab staff's original hypothesis was incorrect.

Another research project added to FeederWatch was the House Finch Eye Disease Survey. This project was initiated by FeederWatch participants, who observed house finches with puffy eyes during the winter months. Since then, participants have noted how the disease, identified as conjunctivitis, has spread throughout North America's house finch population, causing their numbers to decline. Citizen scientists have proven to be an integral part of the scientific research team, documenting a serious population decline that could help in the understanding of disease outbreaks in other animal populations.

What is particularly interesting about this phase of the project is the number of questions staff received about the experimental process. Many of these queries focused on hypotheses that participants were developing to help explain their results. This kind of activity showed that not only were participants fully engaged in the project, but they were also taking scientific inquiry to the next level. They were using scientific methods and applying them appropriately to answer their research questions. As a result, participants were learning about science in the context of real scientific research.

Citizen scientists are becoming indispensable to the research efforts of the Cornell Lab of Ornithology. They are contributing to scientific knowledge about ecology and bird-feeding patterns in their regions. In fact, their findings have been included in articles published in peer-reviewed journals.

"We are not just being nice in letting the public participate in these projects," says Bonney. "Their scientific data are extremely important. Increasingly, the scientific community is depending on this work to further our understanding of North American birds."
(Adapted from Citizen Science at the Cornell Lab of Ornithology, In Exemplary Science in Informal Education Settings, *edited by Robert E. Yager and John Falk. Arlington, Virginia: NSTA Press, 2008.)*
[END OF CASE STUDY 2-1]

This project is a powerful illustration of how an informal experience can provide rich and meaningful opportunities for people to participate in and learn about science. With some guidance from staff, the participants used the tools of science as they learned the practices, goals, and habits of mind of the culture of science. Similarly, the scientific community responded to participants, modifying their project design as a result of feedback and continued interest in the project.

For example, staff added the Seed Preference Test because participants were looking for a new challenge. Through observation over a long period of time, the citizen scientists documented that a hypothesis developed by lab staff was inaccurate. As is done in the scientific community, their findings were shared in articles published in peer-reviewed journals. Through this fruitful collaboration, the relationship between scientists and citizen scientists evolved, resulting in all members contributing and gaining valuable scientific knowledge.

A sustained citizen science experience like Project FeederWatch provides an ideal opportunity for science novices to become familiar with the process and culture of science and even to become engaged participants in the scientific enterprise. Short-term or one-time informal science education experiences will be more challenged to acquaint learners with the culture of science in the fullest sense. Nonetheless, it is still possible to portray the social, lived, and dynamic aspects of science as part of a science exhibition and short programming.[2]

WHAT IS SCIENCE LEARNING?

Research on learning science makes clear that learning science involves development of a broad array of interests, attitudes, knowledge, and competencies. Clearly, learning "just the facts" or learning how to design simple experiments is not sufficient. In order to capture the multifaceted nature of science learning, we adopt the "strands of science learning" framework that was developed in *Learning Science in Informal Environments* that articulates the science-specific capabilities supported by informal environments. This framework builds on a four strand model developed to capture what it means to learn science in school settings.[3] The two additional strands incorporated for learning in informal environments, Strands 1 and 6 below, reflect the special commitment to interest and sustained engagement that is a hallmark of informal settings. The strands provide a framework for thinking about elements of scientific knowledge and practice.

An important aspect of the strands is that they are intertwined, much like the strands of a rope. Research suggests that each strand supports the others, so that progress along one strand promotes progress in the others. It is important to note that, although the strands reflect conceptualizations developed in research, they have not yet been tested empirically. Nonetheless, they provide a useful framework to help educators, exhibit designers, and other practitioners in the informal science education community articulate learning outcomes as they develop programs, activities, exhibits, and events.

Strand 1: Sparking Interest and Excitement

This strand focuses on the motivation to learn science, emotional engagement, curiosity, and willingness to persevere through complicated scientific ideas and procedures over time, all of which are important for learning science.[4] Learners in informal settings experience excitement, interest, and motivation to learn about phenomena in the natural and physical world. Interest includes the excitement, wonder, and surprise that learners may experience. Recent research shows that the emotions associated with interest are a major factor in thinking and learning, helping people learn as well as helping them retain and remember.[5] Engagement can trigger motivation, which leads a learner to seek out additional ways to learn more about a topic.

For example, after a field trip to the local planetarium, young people could become so excited that they decide to join a local astronomy club. In that setting, not only will they learn more about this topic, but they also will meet other people with similar interests.

Strand 2: Understanding Scientific Content and Knowledge

This strand includes knowing, using, and interpreting scientific explanations of the natural world. Learners in informal environments come to generate, understand, remember, and use concepts, explanations, arguments, models, and facts related to science. Learners also must understand interrelations among central scientific concepts and use them to build and critique scientific arguments. While this strand includes what is usually categorized as content, it focuses on concepts and the link between them rather than on discrete facts. It also includes the ability to use this knowledge in one's own life.

For example, after watching a large format IMAX movie about the Galapagos Islands, viewers could be challenged to apply what they learned about natural selection to another environment. After noticing a particular species in that environment, the learner could hypothesize about how a naturally occurring variation led to the organism's suitability to the environment.

Strand 3: Engaging in Scientific Reasoning

This strand encompasses the knowledge and skills needed to reason about evidence and to design and analyze investigations. It includes evaluating evidence and constructing and defending arguments based on evidence. The strand also includes recognizing when there is insufficient evidence to draw a conclusion and determining what kind of additional data are needed. Many informal environments provide learners with opportunities to manipulate, test, explore, predict, question, observe, and make sense of the natural and physical world. In fact, most science centers are built around the concept of exploration. Visitors are not given a correct scientific explanation of a natural phenomenon. Rather, they are presented with a phenomenon and then led through a process of asking questions and arriving at their own answers (which may then be verified against current scientific explanations).

The generation and explanation of evidence is at the core of scientific practice; scientists are constantly refining theories and constructing new models based on observations and experimental data. Understanding the connections, similarities, and differences between the ways people evaluate evidence in their daily lives and the practice of science is also part of this strand (e.g., looking at nutrition labels to decide which food items to purchase, understanding the impact of individual and collective decisions related to the environment, diagnosing and addressing personal health issues, diagnosing and testing different possible causes of a broken appliance).

On a small scale, visitors to science museums have an opportunity to engage in scientific reasoning. In these settings, visitors can interact with phenomena, see what happens, and then develop their own explanations for what they just experienced. For example, after experimenting with different objects to see which ones float and which ones sink, visitors can see that shape is just as important a variable as weight in determining buoyancy.

Through trial and error and by asking questions, people can begin to develop a deeper understanding of the world. The process of asking questions and then determining ways to answer them is often the way that people of all ages learn new ideas. This process can take place in many settings, including the home, a community center, a museum, a lecture, or an informal event such as a Science Café.

Strand 4: Reflecting on Science

The practice of science is a dynamic process, based on the continual evaluation of new evidence and the reassessment of old ideas. In this way, scientists are constantly modifying their view of the world. Learners in informal environments reflect on science as a way of knowing; on processes, concepts, and institutions of science; and on their process of learning about phenomena. This strand also includes an appreciation of how the thinking of scientists and scientific communities changes over time as well as the learners' sense of how his or her own thinking changes.

Research shows that, in general, people do not have a very good understanding of the nature of science and how scientific knowledge accumulates and advances.[6] This limited understanding may be due, in part, to a lack of exposure to opportunities to learn about how scientific knowledge is constructed.[7] It is also the case that simply carrying out scientific investigations does not automatically lead to an understanding of the nature of science. Instead, experiences must be designed to communicate this explicitly.

Informal learning environments and programming are well suited to providing opportunities for people to experience some of the excitement of participation in a process that is constantly open to revision. Developing an understanding of how scientific knowledge evolves can be conveyed in museums and by media through the creative reconstruction of the history of scientific ideas and the depiction of contemporary advances. Also compelling are the human stories behind great scientific discoveries. Such scientists as Galileo Galilei, Benjamin Franklin, Charles Darwin, Marie Curie, James Watson and Francis Crick, and Barbara McClintock are just a few people whose stories provide examples of how scientific ideas evolve.

The nature of science can also be reflected in documentary-style entertainment shows. For example, Myth Busters investigates assumptions about the nature of particular phenomena and Crime Scene Investigations (CSI) depicts evidence as sometimes fragile and temporary in nature.

Strand 5: Using the Tools and Language of Science

The myth of science as a solitary endeavor is misleading. Science is a social process, in which people with knowledge of the language, tools, and core values of the community come together to achieve a greater understanding of the world. The story of how the human genome was mapped is a good example of how scientists with different areas of expertise came together to accomplish a Herculean task that no single scientist could have completed on his or her own. Even small research projects are often tackled by teams of researchers.

Through participation in informal environments, nonscientists can develop a greater appreciation of how scientists work together and the specialized language and tools they have developed. In turn, learners also can refine their own mastery of the language and tools of science. For examples, kids participating in a camp about forensic science come together as a community to solve a particular problem. Using the tools of science, such as chemical tests to identify a substance found at the crime scene, students become more familiar with the means by which scientists work on their research problems.

By engaging in scientific activities, participants also develop greater facility with the language of scientists; terms like *hypothesis*, *experiment*, and *control* begin to appear naturally in their discussion of what they are learning. In these ways, nonscientists begin to gain entrée into the culture of the scientific community.

Strand 6: Identifying with the Scientific Enterprise

Through experiences in informal environments, some people may start to change the way they think about themselves and their relationship to science. They think about themselves as science learners and develop an identity as someone who knows about, uses, and sometimes contributes to science. When a transformation such as this one takes place, young people may begin to think seriously about a career in a health field, in an engineering firm, or in a research laboratory.

Older adults, who have more leisure time after retirement, may take up hobbies that help give them a new identity at this time of their lives. For example, in addition to spending many hours outside cultivating his beds, an amateur gardener also may pursue another passion, such as growing orchids in a greenhouse. To become more knowledgeable, he or she could seek out information in books, online, or at the local botanical garden club. After becoming somewhat of an expert on orchids, the gardener may be asked to talk to senior citizens at an intergenerational center about his hobby, or become a volunteer docent or gardener at a local botanic garden or park. At this point, the gardener has assumed a new identity—as an expert in the field and as a teacher. Changing individual perspectives about science over the life span is a far-reaching goal of informal learning experiences.

Sustaining existing science-related identities may be more common than creating new ones. For example, in one study, visitors to the California Science Center already expressed a strong sense of connection to science, and their visit reinforced their self-image as someone with interest in or connections to science.

Using the Strands Framework

The strands are statements about what learners do when they learn science, reflecting the practical as well as the more abstract, conceptual, and reflective aspects of science learning. The strands also represent important outcomes of science learning. That is, they encompass the knowledge, skills, attitudes, and habits of mind that learners who are fully proficient in science demonstrate. The strands serve as an important resource for guiding the design of informal learning experiences and especially for articulating desired outcomes for learners. Throughout this book, we return to the strands as a way to highlight the learning described in the numerous case studies.

* * * *

Learning science is a multifaceted endeavor. It involves developing positive science-related attitudes, emotions, and identities; learning science practices; appreciating the social and historical context of science; and understanding scientific explanations of the natural world.

Informal environments have often been viewed as particularly important for developing learners' positive science-specific interests, attitudes, and identities

Designers and educators can realize these goals and make science more accessible to people of all ages when they portray it as a social, lived experience relevant to the lives of the learners. Project FeederWatch is an example of such a project. Participants became part of a community of scientists and made their contributions while engaging in science in a familiar context.

As a way to assess science learning in this new framework, this chapter introduced a strands framework The strands provide a way to describe learning outcomes specifying the content, skills, and ideas people are striving to master in informal science settings.

In the next chapter, we look closely at strategies designers can use to make science more accessible to a range of participants. These include interactivity and the importance of presenting information in multiple ways to reflect the needs and interests of a wide range of learners. To further validate these ideas, they are presented in the context of additional research about how people learn.

Things to Try

To apply the ideas presented in this chapter to informal settings, consider the following.

- *What is the culture of your community?* This chapter explores the practices, values, and language that are part of culture. With these ideas in mind, bring together your staff to discuss what elements make up the culture of your environment. Are these elements clear to most people? Do most people buy into them? Do these elements attract visitors, keep them away, or both?
- *Complete an informal survey of your setting as a way to better understand those who visit.* Staff at the Cornell Lab of Ornithology modified their program based on information they learned through surveying participants. Consider surveying participants in your program to learn more about their preferences and what could be modified in your setting to expedite learning.
- *Think about how the strands may apply to your setting.* In this chapter, we introduced the strands as a model that can be used to describe learning outcomes. Consider how the strands may be applied to the learning that takes place in your setting. Do your current offerings encompass all of the strands? Which strands are covered most frequently? Are there any strands that are rarely touched on? Are there any strands that seem particularly important for your setting, but have not been programmed towards?
- *Discuss evaluation data with an outside consultant.* Reviewing evaluation data with an outside expert may help you see the information with fresh eyes. The consultant also may have good ideas of how to use the data more effectively.
- *Involve local learning researchers or educators.* Make use of other resources available in your community to discuss learning and learning outcomes. You could create an advisory group of knowledgeable experts.
- *Join online communities of peers.* There are a variety of listservs and blogs that provide informal science educators with connections and opportunities to discuss learning with peers (e.g. ISEN-L for science museums and science centers).

For Further Reading

Latour, B., and Woolgar, S. (1986). *Laboratory life: The social construction of scientific facts.* Princeton, NJ: Princeton University Press.

National Research Council (1999). Executive Summary in *How People Learn* (pp. xi-xvii). Committee on Developments in the Science of Learning. Division of Behavioral and Social Sciences and Education. Washington, DC: The National Academies Press.

National Research Council (2009). Introduction. Chapters 1 and 2 in Committee on Learning Science in Informal Environments, *Learning Science in Informal Environments: People, Places, and Pursuits.* P. Bell, B. Lewenstein, A.W. Shouse, and M.A. Feder (Eds.). Center for Education, Division of Behavioral Sciences and Social Science and Education. Washington, DC: The National Academies Press.

National Research Council (2008). Four strands of science learning. Chapter 2 in Committee on Science Learning, Kindergarten Through Eighth Grade, *Ready, Set, Science!* S. Michaels, A.W. Shouse, and H.A. Schweingruber (Eds.). Center for Education, Division of Behavioral and Social Sciences and Education. Washington, DC: The National Academies Press.

National Research Council (2007). Goals for science education. Chapter 2 in Committee on Science Learning, Kindergarten Through Eighth Grade, *Taking Science to School.* R.A. Duschl, H.A. Schweingruber, and A.W. Shouse (Eds.). Center for Education, Division of Behavioral and Social Sciences and Education. Washington, DC: The National Academies Press.

Yager, R.E., and Falk, J. (Eds.). (2008). *Exemplary science in informal education settings: Standards-based success stories.* Arlington, VA: NSTA Press.

Web Resources

Center for the Advancement of Informal Science Education (CAISE): http://caise.insci.org/

Citizen Science: www.citizenscience.org

Part II
Designing Experiences to Promote Science Learning

3
Designing for Science Learning: Basic Principles

After visiting a physics exhibit illustrating objects floating on a stream of air, a 13-year-old boy noted the following:

> *Oh, yeah. I was like, oh, I didn't know that. I didn't know it could stay up for so long. I thought eventually it would just die down and the weight would overcome the air pressure and stuff. But it just kept on floating. It was pretty cool (Tisdal, 2004, p. 28).*

An adult visitor to the *Search for Life* exhibit at the New York Hall of Science was very excited after experiencing it:

> *I think the water exhibit is really brilliant. I can read something in a paragraph and not really have a sense of how much water 16 gallons is. It was just beautifully illustrated and really surprising. I had no idea that that much water is in our body. I think the [New York Hall of Science staff] do a great job of taking abstract contents and making it concrete so you can touch it and see it. That's why I like to bring my kids. You're going to absorb something somehow, even if you're not really trying at all (Korn, 2006, p. 17).*

After visiting a conservatory, a young man said the following:

> *(What did you like most about the Conservatory?) "One place that I particularly liked and was pleased with was the Plant Lab because it showed me the way plants come to form life and microscopes show you the different shapes of the seeds, the leaves, the roots—so many things that I didn't know before . . . . I came here and many of them refreshed my memory of when I was a child and took classes at school" (male, age 28; translated from Spanish) (p. 5; p. 139 from Learning Science in Informal Environments).*

Three different informal science experiences on different topics, but each evoked a powerful response from participants, and each resulted in some learning. Success of informal experiences like these is not accidental. Informal science educators typically develop such experiences over time and make gradual improvements based on how learners respond. Not surprisingly, the principles for design that emerge from the work of seasoned informal educators align in many ways with findings from research on learning.

INSIGHTS FROM RESEARCH ON LEARNING

Studies of experts and novices provide insight into what it means to have deep and flexible understanding. Experts in a particular domain are people who have deep, richly interconnected ideas about the world. They are not just good thinkers or really smart. Nor are novices poor thinkers or not smart. Rather, experts have knowledge in a specific domain—such as chess, waiting tables, chemistry, or tennis—and are not generalists. However, experts do not just know "a bunch of facts." In fact, having expertise in a topic means that knowledge is organized into coherent frameworks, and the learner understands the interrelationship between facts and can distinguish which ideas are most central. This kind of deep but organized understanding allows for greater flexibility in learning and facilitates application across multiple contexts.

Research has documented how development of expertise can begin in childhood through informal interaction with family members, media sources, and unique educational experiences[1]. In fact, from early childhood onward, humans develop intuitive ideas about the world, bringing prior knowledge to nearly all learning endeavors. Children and adults explain and hear explanations from others about why the moon is sometimes invisible, how the seasons work, and why things fall, bounce, break, or bend. Interestingly, these ideas or assumptions about how the world works develop without tutoring, and people are often unaware of them. Yet they often influence behavior and come into play during intentional acts of learning and education.

Thus, a major implication for thinking about informal science learning is that what learners understand about the world is perhaps as important as what one wishes for them to learn through a particular experience. Accordingly, efforts to educate should focus on helping learners become aware of and express their own ideas, giving them new information and models that can build on or challenge their intuitive ideas.

Another important feature of experiences that support learning is providing prompts that guide individuals to reflect on their own thinking. This ability to reflect on and monitor one's own thinking, termed "metacognition," is a hallmark of expertise. Metacognition, like expertise, is domain-specific. That is, a particular metacognitive strategy that works in a particular activity (e.g., predicting outcomes, taking notes) may not work in others. However, metacognition is not exclusive to experts; it can be supported and taught. Thus, even for young children and older novices engaged in a new domain or topic of interest, metacognition can be an important means of controlling their own learning.[2] Accordingly, as a means of directing and promoting learning, metacognition may have special importance in informal settings, in which learning is self-paced and frequently not facilitated by an expert teacher or facilitator.

STRATEGIES FOR PUTTING RESEARCH INTO PRACTICE

These facets of learning—the development of expertise, the role of intuitive ideas and prior knowledge in gaining deep understanding, and the ability to reflect on one's own thinking—can be put to use in informal settings to build deeper, more flexible understanding. One way this is accomplished is by creating informal environments that juxtapose the learners' understanding of a natural phenomenon with the formal disciplinary ideas that explain it. This often includes illustrating a surprising or typically hidden aspect of the phenomenon and prompting the learner to reflect on what it means. This approach is intended to help learners examine their own understanding and work toward revising it so that it more closely resembles current scientific understanding.

Another strategy that can aid flexible learning is providing multiple ways for learners to engage with concepts, practices, and phenomena in a particular setting. This strategy reflects the finding that knowledge presented in a variety of contexts is more likely to support flexible transfer of knowledge. For example, in museum settings there is evidence that interpretive materials, such as labels, signs, and audio guides, are more effective in increasing knowledge and understanding than simply interacting with an object or natural phenomenon.[3] Similarly, in more extended experiences, such as those provided by programs, it can be beneficial to provide learners with multiple opportunities to learn about a topic, such as through background reading, presentations, discussions with experts, and direct investigations.

A third strategy identified by researchers and experienced designers is *interactivity*. In her book, *Planning People for Exhibits*, Kathleen McLean defined interactivity as follows: "The visitor acts upon the exhibit, and the exhibit does something that acts upon the visitor." Interactive experiences offer rich opportunities for provoking learners to recognize and reflect on their current ideas. They also allow learners to pursue the questions that might be generated as a result.

There are many different kinds of interactive experiences. Some involve touching or engaging with objects or live animals. Others involve turning knobs, pushing levers, spinning wheels, or doing other manipulations to create an event or see an answer. More extensive interaction might include carrying out a full-blown scientific investigation. In the case of media such as television, the learner may watch others carry out the interactive component, such as doing the steps in an investigation or engaging with an animal.

Interacting directly with materials appears to have particular value. Powerful learning takes place when an individual is able to find out for himself or herself that by correctly connecting wires to a battery, the bulb will light up, or by touching two different kinds of rocks, it is possible to "feel" the difference between them and then classify them accordingly. Making interactive experiences accessible to a wide range of audiences is a distinctive feature of museums, science and nature centers, and other informal science venues.

LEARNING FROM INTERACTIVE EXPERIENCES

There is evidence that interactive experiences support learning across the six strands as well as reflect a concrete way to put the research about learning to work. Such experiences seem to spark interest and maintain learners' engagement while also increasing knowledge and providing opportunities for reasoning. For example, one such exhibit was designed to help visitors understand the form and function of the human skeleton. The exhibit consisted of a stationary bicycle that a visitor could ride next to a large reflecting pane of glass. When the visitor pedaled the bicycle, the exhibit was arranged so that an image of a moving skeleton appeared inside the pedaling person's reflection. The movements of the legs and skeleton attracted the visitor's attention to the role and structure of the lower part of the human skeleton.

According to museum researcher Jack Guichard, the skeleton exhibit experience seemed to transform children's understanding of the skeleton, knowledge related to Strand 2, Understanding Scientific Knowledge. After the cycling experience, children ages 6-7 were given an outline of the human body and asked to "draw the skeleton inside the silhouette." Of the 93 children in the sample, 96 percent correctly drew skeletons whose bones began or ended at the joints of the body; this result was in sharp contrast to the understanding shown by a sample of children of a similar age who did not experience the exhibit; only 3 percent of this group could draw a skeleton correctly. Even more impressively, the children's understanding persisted over time, with 92 percent retaining the idea of bones extending between places where the body bends 8 months after their museum visit. During that time, the children had not received additional schooling, practice, or warning that they would be tested.

Interactive experiences also support Strand 3, Engaging in Scientific Reasoning, although the most sophisticated kinds of reasoning are more difficult to support in short-term experiences. In a study of eight interactive exhibits from three different science centers, Randol (2005) found that the majority could be categorized as "do and see" activities. That is, visitors manipulate the exhibit to explore its capabilities and observe what happens as a result. Through their actions, the visitors engage in many behaviors associated with inquiry, including turning a dial or rolling a wheel, observing what happens, collecting data, and describing results. More sophisticated elements of scientific reasoning have also been observed, such as interpretation of the observed reactions, connecting them to prior experience, predicting outcomes of additional manipulations, and posing further questions. However, in museum settings, these occur less often than simple observation and description.

It appears, too, that providing opportunities for active engagement draws more people to an exhibit (Strand 1). Researcher John Koran and his colleagues found that simply removing the plexiglass cover from an exhibit case of seashells increased the number of visitors who stopped there and the amount of time they spent, even though only 38 percent of those who stopped actually picked up a shell.

Even in institutions with live animals, visitors seek out interactivity. In a study designed by Alexander Goldowsky, visitors were divided into two groups to compare two different learning experiences associated with an exhibit on penguins. The control group went to a typical aquarium exhibit, where they observed live penguins in their natural habitat. The experimental group went to a similar exhibit with an interactive component added—a device that allowed participants to move a light beam across the bottom of the pool. Attracted by the light, the penguins would chase it. After reviewing videotapes for 301 visitor groups (756 individuals), Goldowsky found that those who interacted with penguins were significantly more engaged by the exhibit and more likely to discuss the behavior of the penguins.

Although interactivity has many benefits for learning, it should be used strategically to further the goals of the experience. In fact, research conducted at the Exploratorium in San Francisco reveals that more interactive features are not necessarily better. In one study, museum developers created three different versions of an exhibit called *Glowing Worms.* One was highly interactive (with changeable lighting, focus, and dish location) with live specimens; a second was less interactive (with changeable lighting and focus) with live specimens; and a third was noninteractive (with prerecorded video) with no live specimens. The results of the study showed that visitors who saw one of the two interactive exhibits with live specimens stayed longer, enjoyed the exhibit more, and were able to reconstruct more relevant details of their experience than those who saw the noninteractive exhibit. Yet the researchers found no significant differences between the experiences of the visitors at the less interactive exhibit than at the more interactive one.

These results suggest that adding more features does not necessarily enhance the experience. Extrapolating from this study, Exploratorium staff noted that sometimes too many interactive features can lead to misunderstandings or cause visitors to feel overwhelmed. In fact, these researchers think that there may be an optimal degree of interactivity, which results in a satisfying learning experience for the majority of participants.

[CASE STUDY 3-1]

The following case study of a long-term exhibit called *Cell Lab* at the Science Museum of Minnesota illustrates how these strategies--juxtaposing different ideas to spur reflection, presenting multiple ways to engage with concepts, and interactivity--can prompt learning. Divided into a series of stations, *Cell Lab* offers visitors the opportunity to use real laboratory equipment to conduct short experiments as a way to learn more about cell biology, genetics, microbiology, and enzymes. The opportunity to have such an authentic, or real-world, experience is one of the hallmarks of informal learning environments.

Cell Lab: An Opportunity to Interact with Scientific Instruments

For most people, the tools and practices of science are a mystery. In their daily lives, they do not have opportunities to see laboratory equipment and materials such as bacteria cultures, centrifuges, or even microscopes, let alone actually use them as part of a scientific investigation. However, at the Science Museum of Minnesota's (SMM) *Cell Lab*, museum visitors engage in wet-lab biology activities using real scientific tools and techniques.

Cell Lab consists of a series of eight wet-lab biology activity benches. Each activity is equipped with an online Lab Companion, which introduces the investigation; gives instructions on how to use the instruments, tools, and materials; and leads participants step by step through the procedure. Often the Lab Companion provides supplemental information to enhance the experience. To provide additional assistance, museum volunteers and Lab Crew members—high school juniors and seniors who work in the *Cell Lab*—are available to answer questions.

Cell Lab investigations vary from station to station. At one bench, visitors use toothpicks to scrape cells from the inside of their checks, fix the cells to a slide, stain the cells, and look at the cells under a microscope. The Lab Companion allows further investigation about the structure of cheek cells and the variations they may have noticed.

At another station, Testing Antimicrobials, visitors make a hypothesis about which type of antibacterial cleaner—hand soap, bleach, or sanitizers—most effectively kill a common bacterium, *Bacillus megaterium*. Participants test their hypotheses by using a fluorescent assay to expose bacteria to each agent. If bacteria are still present, they will glow green. If the agent killed the bacteria, the sample does not glow. This activity allows participants to test their hypotheses and see for themselves the impact of cleaning products on bacteria.

To make the experience as safe and authentic as possible, everyone entering *Cell Lab* must put on a lab coat, goggles, and gloves. This laboratory uniform protects the participants, keeps biological sample bacterial contamination to a minimum, and puts the museum visitor into the proper frame of mind. "The lab coat, goggles, and gloves are really a lab uniform, which becomes part of the experience. Our visitors really enjoy dressing as scientists do," says Laurie Fink, director of human biology at the museum.

Visitors' Responses to Cell Lab

Cell Lab has been open for almost 10 years and was the first wet-lab experience created for the public. It also has proven to be a popular attraction at the museum. Over the past 10 years, evaluations have provided the museum with data about who visits *Cell Lab*, what activities they engage in at the different benches, and what their overall impressions of the experience have been. According to a summative evaluation conducted by Randi Korn & Associates, most of the visitors have been small groups of adults and children (often a caregiver and a child), and they spend an average of 15 minutes at each bench. They really enjoy working on the different investigations, with the Cheek Cell bench often selected as one of their favorites. Below are some visitor reactions:

We got to mix all [this] stuff together. (Interviewer: What's fun about that?) "Mixing stuff is fun."

It was spelled out in an easy way, so it was easy for the kids to do on their own.

It was interesting to be able to test some ideas for yourself. Like the anti-bacterial soap and saliva—it didn't tell you what the answer would be, you had to test it for yourself. Then at the end it [the Lab Companion] provided some information. That . . . helped you understand [what] you just did. That [is what] makes these [lab benches] so good—the [combination] of experience and information (male, 43).

The first quote above illustrates a common response of visitors, that is, that one of their goals for an informal experience is active engagement or doing fun things. The second quote is a reminder that visitors may want to explore complex issues, but prefer to do so through experiences that are easily accessible. Parents are often particularly concerned that their children are able to easily participate. The third quote from an adult describes the full spectrum of the experience and again illustrates that learners can be aware of the content and even the underlying design principles of the experience.

Interviews with visitors also reveal that they mostly performed the investigation outlined on the Lab Companion and then talked about what happened. The setup of the benches has been thoughtfully designed to allow for dialogue. "The museum designs these spaces to support social interaction. The benches are arranged so that small groups can do the activities together," explains Kirsten Ellenbogen, director of evaluation and research in learning at the museum. "People can look at each other's experiments or specimens and talk about what they see."

These strategies appear to be working. They have been consistently successful in providing visitors with a rewarding experience. Perhaps a father, visiting with his 11-year-old daughter, best sums up the impact of a visit to *Cell Lab*: "I don't know if I could really speak for the kids, but they always want to come back to the cell ones [Cell Lab benches]. It's my favorite because it's fun to mess around with all this stuff. Do little experiments for yourself rather than watch someone else to do it. We visit all the time and even though the experiment's the same, the kids get just as excited . . . . It's like her own little private laboratory—there are people here to help us and it's not too crowded. . . . I think, for her, it's just the chance to do something you can't anywhere else."

Goals Achieved, Trade-Offs Made

Cell Lab illustrates how the key principles of learning can be incorporated into museum exhibits. The experience itself is interactive; all the stations give visitors an opportunity to use materials to learn something new, such as the structure of cheek cells, or to test their ideas about common household products. Because the labs vary considerably, visitors also are presented with multiple ways to engage with different science concepts.

The strands, too, are reflected in the experiences offered at *Cell Lab*. Having the opportunity to use scientific equipment motivates visitors to explore the different stations (Strand 1). By conducting the experiments, visitors are both adding to their understanding of scientific content and knowledge (Strand 2) and fine-tuning their ability to engage in scientific reasoning by asking questions, developing hypotheses and checking them against experiments, and continuing to push their thinking by asking increasingly complex questions about the world (Strand 3). In this setting, there are numerous opportunities to share ideas, ask questions, and become familiar with the ways that science involves searching for explanations of an event or phenomena (Strand 4). Using the tools of science, such as microscopes and fluorescent assays, and conducting the experiments in the context of a science museum, surrounded by other science learners, visitors become, at least temporarily, part of a community of scientists (Strand 5). And by donning a uniform of science—lab coat, goggles, and clothes—as they engage in scientific experiments, visitors further identify themselves as scientists (Strand 6).

Despite *Cell Lab's* strengths, the exhibit designers note that there is room for improvement. For one, they point out that to ensure that the experience is engaging and accessible to visitors of all ages and backgrounds, certain compromises were made.

"By design, the lab is more of a step-by-step wet-lab experience than an open-ended exploration or investigation," explains Ellenbogen. "This allows visitors to be consistently successful in completing an experiment that they would not typically be able to access."

Fink concurs, noting that having a hands-on experience and a chance to "be a scientist" is very appealing to visitors. In fact, visitors become so engaged that nearly everyone stays and completes at least one investigation, which takes about 15 minutes. Sometimes visitors will complete multiple *Cell Lab* investigations while visiting the museum. Spending that much time at one investigation, let alone multiple ones in the *Cell Lab*, is an extraordinary difference from a typical interaction with an exhibit, which may last only 30 to 60 seconds.

Even that positive outcome has another side. "Some people raise concerns about 'through-put.' In other words, how many people can do an investigation in one day if the experience is 15 minutes instead of 15 seconds," says Ellenbogen. "But it is important to value a range of experiences in a museum, keeping depth and breadth."

From Fink and Ellenbogen's perspective, however, they would like to see the labs accomplish even more. "Right now, we're succeeding at identity development; it's amazing how wearing lab clothes helps visitors see themselves as scientists," says Ellenbogen. "And ownership is built into the experiences; when visitors look into the microscope, they are looking at their own cheek cells. They are highly engaged because they are 'doing science' and seeing themselves in a new way. But there is certainly interest in finding a way to make the experience more open-ended and to touch on more of a range of learning experiences."

Doing that is not easy, however. For one thing, the Lab Companion needs to be updated by a computer programmer, making changes difficult. And there is a fine line between open-ended activities that are challenging but not frustrating, especially for young, inexperienced visitors.

To overcome these obstacles, Fink would like to see museums and science centers collaborate on developing the next generation of wet-lab biology activities. "Sharing activities among museums gives us economies of scale," explains Fink. Other institutions--the Maryland Science Center and the St. Louis Science Center, among others--are experimenting with more flexible lab benches, "so tweaking them for our institution and sharing them is a possibility down the road."

(Adapted from Cell Lab Summative Evaluation, Randi Korn & Associates, and interviews with Laurie Fink and Kirsten Ellenbogen)
[END OF CASE STUDY 3-1]

CHALLENGES OF DESIGNING FOR LEARNING

Ellenbogen and Fink's insights into the strengths and weaknesses of *Cell Lab* point to the issues faced by all exhibit designers. A desire to make the experience challenging but not frustrating and open-ended but with opportunities for success built in are widespread goals throughout the informal science community. Figuring out how to realize these goals was a major goal for Exploratorium designers in their development of Active Prolonged Engagement (APE) exhibits.

Unlike more traditional exhibits, which typically present a phenomenon, provide visitors with an opportunity to observe or interact with it in a prescribed way, and then explain what happened in the label, APE exhibits strive to be more open-ended. Their goal is to give visitors more choices about how to approach and engage with the exhibit, with opportunities for formulating a hypothesis, testing it, learning from the results of their experiments, and performing additional tests.

For example, at an APE exhibit called *Downhill Race,* visitors are asked to race two of six possible disks down parallel tracks to see which one rolls faster. Most visitors hypothesized that the heavier ones would roll the fastest, but, in fact, disks with more of their mass located near the hub roll faster than those with more mass located near the rim. Visitors race disks to figure out which variable, mass or distribution of mass, is more important. Four of the disks have fixed masses, and two have masses whose location can be changed.

Among many visitors, this exhibit evoked excitement and brought out their competitive spirit. Because the participants wanted to win the race, many stuck with it, manipulating the masses until they figured out which rolled the fastest. After successfully completing the race, visitors appeared happy and energized.

Interestingly, evaluators of this exhibit found that visitors who had misconceptions about which disk would roll the fastest were the most engaged by it. This intriguing finding may be attributable to the exhibit's success in making visitors' naïve understanding more salient to them and providing them with the opportunity to explore alternative explanations.

To continue to think about the challenges inherent in exhibit design, we now look at a different kind of exhibit. Called *The Mind,* it, too, was developed at the Exploratorium. The issue facing these designers was how to create an exhibit that explores how the mind—the most elusive and mysterious part of ourselves—functions in different situations.

[CASE STUDY 3-2]

Probing the Depths of *The Mind*: An Exhibit at the Exploratorium

What do you think it would feel like to drink water from a fountain made from a toilet—even if you knew the water was clean? Would you be able to do it, or would it make you feel too uncomfortable? The exhibit *The Mind* explores such issues in a series of 40 interactive experiences.

During the four-year development process, senior exhibit developer Erik Thogersen and his team worked with expert advisers to design experiences that would provide visitors with some insight into how they make decisions or respond to unusual, counterintuitive events.

In determining how to accomplish this goal, Thogersen considered many approaches, including the model pioneered in APE project. But he soon realized that for *The Mind* exhibit, the activities didn't have to be open-ended in the same way the APE exhibits were. One reason is that its purpose was not to foster skill development but to trigger responses that would reveal something about how the mind works. It soon became clear that one approach would not be sufficient; ultimately, the exhibit would use different strategies to elicit the desired responses and learning.

At one station, a visitor looks a screen showing a small patch of skin that has been magnified many times. The visitor is asked to think about an emotionally arousing idea or image. The thoughts trigger an immediate secretion of sweat, which shows up on the screen, presenting a concrete physiological reaction to a cognitive event.

Through trial and error, Thogersen discovered another effective approach—designing interactive exhibits for small groups. Such exhibits, says Thogersen, gave visitors a chance "to prod the minds of others" and were particularly effective if members of a group knew each other.

For example, at a station designed to measure emotional reactions by graphing breathing patterns, visitors are given 12 cards with questions, such as "Name somebody you have a crush on." Two people who don't know each other can ask that question, but the experience is much more powerful when those two people are friends. "If you know the person, you can ask that question in a way that embarrasses them or gets at another emotion," Thogersen points out. "The reaction you get is much stronger—and much more interesting." Even activities that can be done alone, such as experiencing the toilet water fountain or measuring reaction time to sensory stimulus, are more fun when done with a partner.

Interestingly, because of the Exploratorium's development process—exhibits are prototyped and released in groups over a long period of time--Thogersen didn't realize that about half of the activities were designed for more than one person until after the whole exhibit had been completed. "We noticed it more in retrospect," Thogersen admits. "It just kept happening, probably because it turned out to be the best way to explore something as abstract as the mind."

Looking ahead, Thogersen and his colleagues are always thinking about innovative ways to design exhibits that elicit strong responses and bring about learning. The possibility of using computer visualizations to model phenomena that can't be shown, such as the topography of the San Francisco Bay, is one new intriguing idea. This exhibit is currently on display. To introduce an element of interactivity, visitors are asked to use the cursor to drop a virtual can into the bay and then observe how far and in what direction the currents carry it. Visitors can drop the cans anywhere in the bay to compare currents, or they can drop in a whole flotilla to see how small differences in initial placement eventually bring the cans to very different places.

Aware that there is room for improvement, Thogersen acknowledges that finding new ways to excite visitors is an ongoing challenge. "We're always stretching ourselves out of our comfort zone to push ways to bring about engagement," he says. "We're always looking for ways to show visitors really cool things that can happen."

(Adapted from an interview with Erik Thogersen, senior exhibit developer at the Exploratorium)

[END OF CASE STUDY 3-2]

Like the designers of *Cell Lab*, Thogersen's team also considered ways to spur learning in developing *The Mind* exhibits. In doing so, this exhibit illustrated the effectiveness of displaying concepts in multiple ways and creating a unique set of interactive activities.

While exploring *The Mind,* visitors were stimulated by the variety of options available to them (Strand 1). As they engaged with the exhibit, they were introduced to content knowledge, some of which was probably new to many (Strand 2).

One of the most interesting features of this exhibit is the way it became a social process, largely because many of the stations were designed for two people. What's more, Thogersen noted that the learning was even more powerful if the two people knew each other. The importance of the social nature of learning (Strand 5) is explored in more detail in the next chapter.

Both of these examples illustrate how science museums have incorporated strategies that are supported by research to develop experiences that foster learning. These same strategies can be put to use in programs, which continue over an extended period of time. In fact, experiences that occur over a longer period of time can provide opportunities to encourage learning across the six strands.

The following case study illustrates the learning that occurs in the context of a program. This experience involves teens working with younger children as part of a program at the St. Louis Science Center.

[CASE STUDY 3-3]

Teenage Designers of Learning Spaces: Science Learning Among Kids of All Ages

A group of teens from the St. Louis Science Center surveyed a large room in a homeless shelter. As participants in a program called Teenage Designers of Learning Spaces, the teens were trying to figure out how to transform the space into an area in which young children, residents of the shelter, could learn more about botany.

The teens faced many obstacles. The room was dark, although there was a set of French doors leading to a patio. While they considered growing plants on the patio, they quickly noticed that it was structurally unsound, making it impossible for the kids to go out there. But light *did* shine through from the patio.

Working together, the teens came up with an ingenious idea: They took a clear shoe bag and hung it on the door. In each pocket, they planted a different kind of flower. Then they developed a field guide to explain the parts of plants and what the plants needed to grow.

"The results were amazing," explains Diane Miller, vice president of education at the St. Louis Science Center and director of the project. "Flowers were growing everywhere. Everyone—staff, the teens, and the young residents of the shelter--couldn't believe that what was once a bulb was now a tulip. And they also learned that it's a good thing to get dirty sometimes. You can touch dirt, you can grow plants, be fascinated by it."

No one was prouder of their success in growing plants than the teens themselves. They had been working toward this goal for quite a while. For two years, they had gone to the science center every Saturday during the school year and every day during the summer. They had learned how to conduct scientific investigations and had acquired other skills they would need to work with younger kids and to modify their assigned spaces to get the best results. Their persistence and hard work had paid off. Not only were the spaces now suitable for science, but by the end of the program, most of the teens were headed to college. Before participating, more than half had been D or F students.

What had made the difference? How had this program resulted in such a marked transformation in so many students?

"We broke through the barriers they had to learning," explains Miller. "We figured out that one of the biggest deficits was a lack of experience with the natural world. We set out to fill in those gaps by providing the kids with real-world problems and the opportunity to solve them by working together. When they became comfortable, then they could learn."

For example, most of the teens in the program had never had a pet, so they decided they wanted to purchase a fish tank. But after setting it up, it didn't take long for all the fish to die.

"What happened?" the kids wanted to know. "Why did all the fish die?" While providing guidance, Miller and her colleagues encouraged the teens to find the answers on their own, in any way they could. So they read about the problem in books and on the Internet, and they discovered that many variables—including water temperature, the composition of the water, the diet of the fish, and the amount of waste (ammonia) the fish produce—contribute to the health of the fish population. They collected data about their own fish and uncovered the reason for their demise: the ammonia levels in the tank were too high. By changing the water more frequently, the kids could prevent this from happening again.

"Once students have an experience like this, when they see that they can solve a problem they find compelling, the first major barrier is removed," explains Miller. "From that point, they become interested and motivated to develop the skills they need to become thinkers and problem solvers."

To assess each student's progress, Miller and her colleagues asked students to demonstrate what they had learned, often by asking each student to develop a work product. For example, to explain their thinking about the design of a learning space, students produced detailed drawings illustrating each design element. Then each student gave a presentation about his or her design. "Their presentations were very articulate," says Miller. "They revealed that the kids were not mimicking what someone else had said. They had internalized what they were describing."

But perhaps most significant of all, as a result of these experiences, the teens felt differently about themselves. Not only could they solve problems, they could solve *scientific* problems. "In our culture, if you can do science, then you must be really smart," says Miller. "If you can do science, you can do anything. By uncovering the hidden scientist in each student, their identity changed, from non-learner to learner."

As part of the assessment, one of the teens in the program described this transformation in his own words. "I was misdiagnosed," he concluded. "I was told I was stupid, but if I can teach at a science center, I must be smart."
[END OF CASE STUDY 3-3]

Supporting Learning

The participants in Teenage Designers of Learning Spaces faced a different challenge than visitors to the two museum exhibits described earlier. They lacked experience with the natural world and had not had opportunities to explore natural phenomena and develop rich, intuitive ideas about them.

After defining the reason why learning had been difficult for this group, Miller and her team discovered the way to connect with the teens—by asking them to find a compelling real-world problem, which they worked on together to solve. The problem highlighted in the case study was figuring out why their goldfish died. Rather than telling the teens the answer, Miller encouraged them to find out what happened on their own. They talked among themselves, read books, and surfed the Internet until they learned what fish need to survive and what was lacking in the environment they had created for their fish. As discussed earlier, research has discovered that learning through multiple channels—books, the Internet, and conversation—tends to support flexible transfer of knowledge. This approach also proved to be so empowering that it set the teens on a path to further learning. After learning the tools and vocabulary of science (Strand 4), they were ready to ask more questions and find new ways to answer them.

Although the teens were novices, they used some metacognitive strategies to find the answer to their problem. They took notes as they did their research, which was conducted in different modalities (books and the Internet). Then, through a process of elimination based on acquired knowledge, they determined which variable (an excess amount of ammonia) was causing the fish to die. Once they knew what the problem was, they had no trouble coming up with a solution—changing the water more frequently.

Two factors determined the amount of learning that took place—time and the quality of the teaching available. Programs represent informal learning experiences that take place over a longer period of time; in this case, it was over a period of two years. Miller and her team took advantage of the time they had to work closely with the teens, getting to know them and finding ways to remove barriers to learning. In contrast to short-term experiences, much can be gained if time is used effectively.

LEARNING THROUGH MEDIA

In the previous case studies, we illustrated how the principles of learning were used to design two museum exhibits and one long-term program. Learning through media requires some different design strategies, although the basic learning principles still apply. For example, television producers cannot provide viewers the opportunity to interact directly with actual phenomena. Instead, they need to find another approach to make science experiences come alive. So they opt for the next best thing—showing viewers what scientific investigations look like. For this experience, producers also use the term "interactivity." In fact, they describe their job as "telling a story about science inquiry"; the possibility for interactivity lies in reproducing the process at home, with support potentially available from the Internet.

While interactivity in a direct sense cannot be part of the television experiences, watching the film encourages the learner to replicate the experience. The television producers are then challenged to produce accompanying activities that use the best design principles in informal learning. In fact, these supplementary learning experiences (often supported by an interactive website) have become the norm for television documentaries, IMAX movies, and planetarium shows.

Another goal of television producers is to help viewers at home learn science. In order to be successful at that job, they rely on storytelling devices unique to this medium. The next case study discusses these devices, along with a discussion of the learning that results after elementary school students watch an episode of *DragonflyTV: Going Places in Science*, which is produced by Twin Cities Public Television.

[CASE STUDY 3-4]

DragonflyTV: What Kids Learn and Why They Learn It

DragonflyTV (DFTV) is committed to showing "real kids doing real science," and the show lives up to its promise. To help make the science come alive even more, during seasons V and VI (2005-2007), *DragonflyTV* producers partnered with science centers to show viewers at home how they can be a place to go to help them explore science. For this reason, each episode begins with a science question raised by the student investigators captured on the show. The viewers watch the investigators use the centers' resources to answer it, taking advantage of a link between two informal science platforms. The hope is that the viewers will realize that they, too, can turn to a local science center to answer their questions about science.

A quick look at any of the episodes shows kids having fun and engaging in a full cycle of inquiry. For example, take the episode titled *Balloons.* Two boys go to Explora, a small science center in Albuquerque, New Mexico, where they pose the following question: How large does a hot air balloon need to be in order to stay in the air? Carefully and methodically, the boys go about trying to answer this question. First, they measure the air temperature so that they can be sure that this variable remains the same through all three trials. Then the young scientists build three model balloons of different sizes and test how much weight they can lift by putting pennies in their baskets and launching them. Next, they figure out the volume and weight of each balloon.

At the end of each trial, the boys write down how they performed the calculations. The episode ends with a trip to the Albuquerque Hot Air Balloon Festival, where the investigators did some more science (predicted and measured the capacity of an actual balloon) and then had the ultimate experience—a ride in a real hot air balloon.

What Did Viewers at Home Learn?

The boys onscreen learned that the more passengers a hot air balloon needed to carry, the greater its volume needed to be. This information was made clear to them through the series of investigations they conducted. The viewers at home, however, did not have this opportunity. Despite that limitation, were they able to grasp the main ideas being conveyed during the episode?

Alice Apley, a researcher with RMC Research Corporation, conducted an evaluation to try to answer this question. Working with 174 fourth and fifth graders from a range of socioeconomic and economic backgrounds in Massachusetts and New Mexico, Apley first wanted to find out what the students remembered from the episode. For example, did they understand what the investigation was about? Did they recall the steps of the investigation and the kinds of data that were collected?

After conducting extensive interviews with the students, Apley found that the viewers were able to answer these questions easily: 61 percent could accurately explain the point of the investigation. Some students described it in terms of the size and weight of the balloon: "trying to learn how balloons carry weight, how it can stay in the air," "how the balloons float in the air, how big it has to be to float," and "how they flew and which size flew better." Other students responded in a more general way, saying that the inquiry was about how hot air balloons fly and stay in the air.

An impressive 90 percent of the viewers understood that the child investigators were testing balloons of different sizes and/or weights to determine how well they stayed in the air. For example, one student said that "first they try to put hot air in different size plastic bags, but when the plastic bags melted, they decided to make balloons with the tissue paper. They put baskets and added pennies to see if the balloons would go up still." And 54 percent understood that balloon size was measured in order to calculate volume, which they expressed as follows:

"They measured the tissue paper to see how big the balloon was and how much hot air would go into it."

Overall, almost all the viewers (93 percent) picked up the main point: balloon size must increase to lift more weight. One viewer expressed this idea as follows: "Make the balloon bigger and bigger volume for more passengers."

What Features of the Show Helped Students Learn?

One of the goals of the study was to try to figure out what storytelling devices were the most successful in facilitating learning. Apley considered the pace of the episode, its visual appeal, the presentation of the inquiry question, and the use of graphics.

Students noted that they liked *Balloons* because the segment "did not go too fast." Students also said that they enjoyed the boys' approach to the problem, characterized by their decision to follow a sequence of tests, adding a new variable at each stage. Perhaps one reason viewers liked this approach was that the experimental procedure was repeated several times, giving them an opportunity to participate and watch as the drama unfolds. At the same time, the episode stayed interesting because a new variable was introduced with each new trial. The segment concluded with a recap of the relationship between balloon size and the weight it can lift—a summary device that helped solidify learning.

One lesson to be learned from this study is to pay close attention to the way the inquiry is presented. A clear explanation up front, an interesting question, followed by a logical sequence of investigations, with some repetition to reinforce the main ideas, are storytelling devices that have proven to be effective. A straightforward conclusion, in which the ideas are recapped and summarized, also is helpful. "The trial-and-error approach was engaging for kids," says Apley. "The kids could follow along with each trial, participating in the drama. It made sense to build a balloon, measure it, and then watch it fly."
(Adapted from "Apley, A., Graham, W.J., Frankel, S.L. RMC Research Corporation. (2007). DragonflyTV: Going Places in Science Children's Viewing Study [Season 6])
[END OF CASE STUDY 3-4]

This case shows that while television shows (or films) cannot use true interactivity to support learning, they can be designed in ways that successfully support learning. In the *DragonflyTV* example, the compelling narrative and the viewers' potential ability to imagine themselves in the role of the boys carrying out the investigations kept viewers engaged. The step-by-step unfolding of the investigation probably helped viewers to think actively about what was happening and reflect on the results. One important point: although the specific design options available across different settings and experiences may vary, the underlying principles of how people learn do not change fundamentally.

[Note that Museum 2.0 and the sidebar on cell phones were moved to Chapter 8]

* * * *

Research on learning points to strategies that informal science educators can use to enhance opportunities for learning. In particular, research on expertise development, building on prior knowledge, and metacognition suggests that program and exhibit developers provide multiple ways for learners to engage with concepts, practices, and phenomena in a particular setting and prompt and support participants to interpret their learning experiences in light of relevant prior knowledge, experiences, and interests. In addition, because learners are diverse, bringing to the informal setting a range of interests and motivations, it is important to create an experience that is multifaceted, interactive, and developed in light of science-specific learning goals.

Continuing with our discussion about how research informs practices in informal settings, Chapter 4 focuses on the social aspects of learning. By designing environments that encourage conversation and support mediation among learners, informal science educators can help their visitors gain deeper knowledge from even one experience and enjoy themselves more in the process.

Things to Try

To apply the ideas presented in this chapter to informal settings, consider the following:

- *Think about the balance between interactive and noninteractive learning opportunities in your setting.* Research supports interactivity as a way to engage visitors and audiences with the informal experience and support various modes of learning. Does interactivity support the learning goals of your setting? If so, are there relatively simple, inexpensive ways to make some of these experiences more interactive?
- *Consider how the research discussed in this chapter could help inform program or exhibit design.* For example, are there ways to provide more pathways to learning in your setting? Are prompts, such as labels, signs, and audio guides, available? Are there opportunities to support and encourage learners to extend their learning over time or across settings? Do your experiences invite re-engagement or repeat visits?
- *Build relationships with neighboring venues.* Contact nearby informal learning science environments to discuss common design issues. Is there a way to pool resources to provide visitors with a unique experience that invites them to seek out more in your or other settings? Are connections being made between the current experience and potential future ones? Are there resources for visitors or audiences that summarize all of the local offerings in a comprehensive way? Are there additional resources such as traveling exhibits that could be brought in to augment the offerings of the setting? These strategies can help facilitate science learning across multiple settings.

For Further Reading

Allen, S. (2004). Designs for learning: Studying science museum exhibits that do more than entertain. *Science Education, 88*(Suppl. 1), S17-S33.

Allen, S., and Gutwill, J. (2004). Designing science museum exhibits with multiple interactive features: Five common pitfalls. *Curator, 47*(2), 199-212.

Apley, A. (2006). *DragonflyTV GPS: Going Places in Science study of collaboration between museums and media.* Accessed at www.informalscience.org.

Apley, A. (2007). *DragonflyTV: Going Places in Science children's viewing study*. Accessed at www.informalscience.org.

Falk, J.H., Scott, C., Dierking, L.D., Rennie, L.J. and Cohen Jones, M. (2004). Interactives and visitor learning. *Curator, 47*(2), 171-198.

Falk, J.H., Dierking, L.D., Rennie, L. and Scott, C. (2005). In praise of "both-and" rather than "either-or:" A reply to Harris Shettel. *Curator, 48*(4), 475-477.

Flagg, B. (2006). Review of museum-based segments of *DragonflyTV: Going Places in Science.* Research Report No. 06-013.

Goldowsky, N. (2002*). Lessons from life: Learning from exhibits, animals and interaction in a museum.* UMI#3055856. Unpublished doctoral dissertation, Harvard University.

Guichard, H. (1995). Designing tools to develop the conception of learners. *International Journal of Science Education, 17*(2), 243-253.

Humphrey, T., and Gutwill, J.P. (Eds.). (2005). *Fostering active prolonged engagement: The art of creating APE exhibits*. San Francisco: The Exploratorium.

Koran, J.J., Koran, M.L., and Longino, S.J. (1986). The relationship of age, sex, attraction, and holding power with two types of science exhibits. *Curator, 29*(3), 227-235.

Korn, R. (2006). *Search for life: Summative evaluation.* New York: New York Hall of Science. Available at: http://www.informalscience.org/evaluations/report 151.pdf

Korn, R. (2003). Cell Lab: Summative evaluation. St. Paul: Science Museum of Minnesota. Available at http://www.informalscience.org/evaluations/report_174.pdf.

McLean, K. (1993). *Planning for people in museum exhibitions*. Washington, DC: Association of Science-Technology Centers.

National Science Foundation (2006). *Now showing: Science Museum of Minnesota 'Cell Lab'*. Available at: ttp://www.nsf.gov/news/now_showing/museums/cell_lab.jsp.

National Research Council (2009). Introduction. Chapters 5 and 8 in Committee on Learning Science in Informal Environments, *Learning Science in Informal Environments: People, Places, and Pursuits.* P. Bell, B. Lewenstein, A.W. Shouse, and M.A. Feder (Eds.). Center for Education, Division of Behavioral Sciences and Social Science and Education. Washington, DC: The National Academies Press.

Randol, S.M. (2005). *The nature of inquiry in science centers: Describing and assessing inquiry at exhibits.* Unpublished doctoral dissertation, University of California, Berkeley.

Rockman, S. (1999). Bill Nye family fun calendar of science evaluation. Accessed at http://www.rockman.com/projects/project/Detail.php?id=124.

Shettel, H. (2005). Commentary on Falk, Scott, Dierking, Rennie and Cohen Jones. Interactives and visitor learning. *Curator*, *48*(2).

Tisdal, C., and Perry, D.L. (2004). Going APE! at the Exploratorium. Interim Summative Evaluation Report. Chicago: Selinda Research Associates, Inc. Accessed at www.exploratorium.edu/partner/pdf/APE_Summative_Phase_1.pdf.

Web Resources

Center for the Advancement of Informal Science Education (CAISE): http://caise.insci.org/

DragonflyTV: http://pbskids.org/dragonflytv/

Exploratorium: http://www.exploratorium.edu

Exploratorium in Second Life: http://www.exploratorium.edu/worlds/secondlife/index.html

Informal Science: http://www.informalscience.org/

Liberty Science Center: http://www.lsc.org/

Museum 2.0: http://museumtwo.blogspot.com/

Science Museum of Minnesota: http://www.smm.org/

Smithsonian 2.0: http://smithsonian20.si.edu/

4
Learning with and from Others

One of the first stops at the Exploratorium's *Frogs* exhibit is a large open tank with a small stream, a stony beach, and dense vegetation. At first glance, the tank appears to be empty, causing visitors to pause and wonder why the Exploratorium would put an empty tank on display. But after looking at it for a while, visitors notice that frogs and toads do in fact live inside. It is because these animals are nocturnal that they are so difficult to see during the day.

This process of puzzlement, surprise, and discovery brings about animated conversation. Below is a sample of the reactions of a child and an adult:

Child: I don't see that many frogs there.
Adult: Do you see any at all? I don't see any at all.
Child: I don't see any. I don't see any frogs. Do you see them?
Adult: Does anybody see any frogs?
Child: I don't see any.
Adult: I don't see any frogs.
Child: I don't see them either. Maybe they're hiding.
Adult: No, I don't see any. OK.
Child: I see that one, frogs and toads. Do you see that one?
Adult: Is that a real one?
Child: Yes, that's a real one.

Adult: Looks like a big one, my heaven.
Child: Is that a toad?
Adult: Yes, I guess that would be a toad. It's sort of on dry land.
Child: It's a pretty one. There's only one in there.
Adult: That's true. They're really wonderful to touch. They have this . . .
Child: I only touched a frog once.
Adult: I know, it's kind of slippery.
Child: They're hard to hang on to.
Adult: Yea, they're hard to hang on to. But it's like touching a live, well it is touching a live creature. That ordinarily isn't used to being touched, like a cat or dog.
Child: That's a pretty, that's a beautiful frog.

Reactions like these, showing curiosity, discovery, and personal responses to an informal science experience, are what museum designers are striving for. These responses reflect Strand 1, and they are essential to the learning process.

In this chapter we explore how interaction with other people plays a role in learning. There are converging reasons to look at learning from and with others as a foundational part of science learning. First, individual learning is supported through interaction with more knowledgeable others and through dynamic exchange of ideas and reflection. Second, as highlighted in Strand 5, science itself involves specialized norms for interacting and specialized forms of language. Learning science therefore involves learning those norms and language. Third, people very often participate in informal science learning experiences with other people. Therefore, the experiences should be designed with groups in mind and in a way that capitalizes on opportunities to engage with other people.

Parents, adult caregivers, peers, educators, facilitators, and mentors play critical roles in supporting science learning. There is ample evidence that children and adults reason about issues that are important to them while interacting with other people. Studies of dinner table conversations, visits to the zoo, and other everyday activities have uncovered rich conversations on a myriad of scientific topics and using scientific forms of discourse.[1] Families of all backgrounds engage in everyday conversations about a broad range of topics, including physics, biology, and religion.[2] Through these kinds of interactions, children engage with others in questioning, explaining, making predictions, and evaluating evidence.[3] Thus, in a variety of ways, including family activities and conversation, children may begin to learn about topics that are relevant to science, even when "learning science" is not an explicit goal of the activity.[4]

As an example, consider the case of watching television. Although most people think of watching television as a solitary activity, when adults or older siblings become involved, the activity can become social, conversational—and more productive. A study with 23 3- and 4-year-old white, middle-class children conducted by Robert Reiser and his colleagues focused on the value of adult-facilitated sessions of *Sesame Street*. During the show, the adults in the experimental group intervened and asked the children to name the letters and numbers shown on the screen, while the control group did not have such conversations. Three days later, the children in the experimental group were better able to name the letters and numbers, suggesting that adult involvement can support learning.

In another study, Margaret Haefner and Ellen Wartella, both researchers in communications studies, found that older siblings could help their younger brothers and sisters understand plot elements in educational programming. Through explanations and laughter, "older children did influence the younger children's general evaluations of the program characters." Even though these studies were not on science programming, their results suggest that active engagement during viewing could have a positive impact for science learning as well. Even an intrinsically passive medium such as television can become interactive when a social, conversational element is introduced. Through conversation and questioning, the ideas embedded in a television programs can resonate for young viewers.

Older children and adults also benefit from interaction with others. In group interactions during museum visits, individuals with more knowledge about a particular exhibit may play an important role in facilitating the learning of others by pointing out critical elements or information and by providing input and structure for a more focused discussion of science.[5] In a small study of an exhibit about glass, adults with high prior knowledge and interest in glass tended to discuss how or why something happened or work more often than those with less prior knowledge or interest.[6] In another example from the museum context, visitors' activities at an exhibit were affected by other visitors' behavior, even when the other visitors are strangers. In one study, adult visitors especially were more likely to touch or manipulate an exhibit if they had previously witnessed a person silently modeling these behaviors.[7]

The importance of more knowledgeable others is reflected in the roles of mentors or scientists in many informal experiences. In citizen science experiences, for example, the relationship between the scientists and volunteers is critical to the volunteers' learning. In the citizen science case study in Chapter 2 (the Cornell Lab of Ornithology), volunteers learned so much from their mentors that they developed enough expertise to contribute to scientific journals.

Similar relationships between experts and novices are important elements of many after-school programs. For example, in a program called Service at Salado, middle school students, undergraduate student mentors, and university-based scientists worked together to learn about an urban riverbed habitat through classrooms lessons and service and learning activities. At the end of the program the undergraduate mentors worked with the middle school students on products to benefit the urban riverbed habitat.[8]

In *The Mind* exhibit described in Chapter 3, museum designer Thogersen discovered that social interaction was key to learning about a concept as abstract as the mind. He notes that two friends could "prod the mind of the other," creating a powerful learning experience for both visitors. Similarly, the benches at *Cell Lab* are organized in such a way to encourage dialogue, building on what research has confirmed—learning is enhanced through social interactions and conversation.

CONVERSATIONS AND LANGUAGE

Conversations are a kind of social interaction that has been studied extensively, especially in museums and classrooms. Engaging in conversation and discussion promotes learning as well as providing a window into the thinking of individuals or groups. By listening to what people say, researchers can find out what learners know and understand, what emotions have been evoked by an experience, and what gaps in learning may remain. The importance of discourse in learning is broadly acknowledged across a range of subject areas and settings.[9] In the classroom context, researchers have found that successful science education depends on the learners' involvement in forms of communication and reasoning that models those of scientific communities.[10] There is increasing interest in designing programs, exhibits, and other informal experiences to promote science learning that explicitly support conversation and use of scientific language. See Box 4-1 for discussion of one strategy.

Studying conversation in informal settings poses many challenges. Among the challenges are determining appropriate ways to record conversations (for example, setting up microphones at selected places throughout the setting versus asking visitors to wear microphones), determining where visitors are in the museum while they are talking (some researchers use "trackers" to follow visitors in the study to identify their movements), and obtaining clear recordings in a noisy environment with notoriously poor acoustics. Transcribing the conversations, the next step in the process, is difficult, time-consuming, and expensive.

By far, however, the most demanding part of such a study is interpreting what the conversations mean. Sue Allen, a researcher at the Exploratorium, points out that not only is the museum environment dense and complex, but also many variables can influence what visitors say and don't say. For example, visitors come from different backgrounds and bring to the visit a range of experiences and varying levels of interest in science as well as diverse attitudes, expectations, group dynamics, and even energy and comfort levels. In addition, aspects of the physical space (lighting and ambient noise, for example), as well as issues related to the design of the exhibit—height, coloration, physical accessibility, interface, display style, label content, and tone—must be taken into account.

Despite these challenges, Allen and other researchers have been able to identify features of conversation that reveal specific kinds of learning. These studies provide insight into participants' thinking and ideas about how informal experiences can be designed to facilitate learning with and from others.

[CASE STUDY 4-1]

Listening to Conversations at the *Frogs* Exhibit

When Exploratorium staff members set out to develop and design the *Frogs* exhibit, they had multiple goals. They wanted to create something beautiful, intriguing, and informative for their diverse audience while also presenting scientific information and engendering respect and appreciation for the animals.

To realize these goals, the exhibit turned out to be quite extensive. It included an introductory area that explained the development of frogs and toads, a section on eating (and being eaten), frog and toad calls, amphibian anatomy, a close-up observation area, a section showing adaptations, a section discussing the declining status of frogs worldwide, and a section on frog locomotion.

In order to explore these varied topics, the final exhibit included an unusually diverse range of exhibit types. Among the offerings were interactive exhibits (10); terrariums of live frogs and toads (23); cases of cultural artifacts (5); samples of maps, excerpts from children's books, and examples of frog folklore (18); cases of organic materials (3); videos of frogs (3); windows to the Frog Lab, where frogs could rest and breed (2); and an immersion experience of sitting on a "back porch" at night listening to the calls of frogs.

In trying to understand the kinds of conversations people (groups of two were selected for the study) engaged in at the exhibit and what they revealed about learning, a coding system was developed that distinguished five overall categories for talk: perceptual, conceptual, connecting, strategic, and affective.

Perceptual Talk

Allen coined the term *perceptual talk* to describe the process of identifying and sharing what is significant in a complex environment. She defined four subcategories that describe the process of sharing one's immediate experience in more detail:

- *Identification,* or pointing out something to attend to, such as an object or interesting part of the exhibit, reflected in such comments as, "Oh, look at this guy," and "There's a tube."
- *Naming,* or pointing out something to attend to, such as an object or an interesting part of the exhibit, such as, "Oh, it's a Golden Frog."
- *Pointing out a feature,* or making note of some concrete aspect or property of the exhibit, such as, "Check out the bump on his head," or "That's loud, huh?"
- *Quoting from a label,* such as "Let's see what it says. The difference between frogs and toads . . . toads live where it's drier. Frogs live in wet places. There's a toad in that tank, supposed to be."

The results showed that perceptual talk occurs at 70 percent of the exhibit types, making it the type of conversation heard most frequently.

Conceptual Talk

This category covered simple inferences, such as a single interpretive statement or interpretation ("They eat mice," someone said after seeing a jar containing a mouse in a display of frog food), and complex inferences, which refer to any generalizations about the exhibit or hypotheses about the relationship between objects or properties.

The Frogs and Toads tank, described in the beginning of the chapter, turned out to evoke complex inferences most frequently. In addition to searching for the 17 kinds of nocturnal marine toads living in the tank, visitors developed hypotheses for why the toads might not be visible. The following short conversation is an example of visitors' thinking:

Visitor 1: Would they bury themselves?
Visitor 2: Perhaps, yeah, or they may really be camouflaged, too.
Visitor 1: Maybe it's just showing where they live.
Visitor 2: Something must be under here because, see, the water is moving.

Exchanges like these indicate that the mystery surrounding the tank was worthy of conversation and discussion. According to Allen, this feature of the exhibit "facilitated learning in an unexpected but fruitful manner." In a surprising result, Allen found that conceptual talk, which is the kind of evidence of learning that museum professionals want to see but often don't, occurred at 56 percent of the exhibit types.

Also in this category is metacognition, the ability to reflect on one's own knowledge and learning. These kinds of comments reveal what visitors notice as they peruse an exhibit and how their observations confirm or contradict what they already know. At the *Frogs* exhibit, a video piece called "Mealtime," which showed frogs catching and eating their food, caught visitors by surprise, leading to such comments as, "I never would have believed . . . " or "I didn't realize they got them with their tongue."

Connecting Talk

This kind of talk refers to connections made between an exhibit and a personal association ("Yeah, my grandmother loves to collect stuff with frogs all over it"); an exhibit and prior knowledge ("In Florida, the dogs eat poisonous toads and die"); and between two exhibits ("That's what I said. It eats anything as long as it fits in its mouth," referring to the label from a previous part of the exhibit.).

At this particular exhibit, the most frequent personal connection was made after viewing a graphic representation of a leaf from the children's book *Frog and Toad Are Friends.* Visitors said the following:

"Oh, this is, oh, look, it's Frog and Toad. I remember that one. That was your favorite book when you were little."
V1: 'Cos here's the story, Mom. I used to have these books a lot.
V2: Oh you're right. Oh, you're exactly right. That's Toad and Frog.

Strategic Talk

This category encompasses two areas—how to use and manipulate the exhibit ("Okay go down to the water . . . and then go towards the back. See that little leafy type thing?" [which was said when someone was searching for the leaf frog.]) and expressions of evaluation of one's own or partner's performance ("I don't think I did a very good job of it.").

Although this kind of talk was heard relatively infrequently throughout the exhibit, the one exception was visitors' responses to an audio-based multimedia exhibit of frog calls. After listening, the visitors could record their imitations of the calls. Visitors comments included much commentary on how they did: "You have to do it before the red line disappears or it doesn't record," and "This was right, except I made it too long."

Allen proposed that several features of this particular exhibit probably accounted for the high frequency of evaluative talk: high overall appeal of the exhibit, a challenging interface to work with, and computer-generated graphs that supported visitors' efforts to visually compare their vocalizations with the standard frog calls.

Affective Talk

This kind of talk refers to emotional responses, such as pleasure, displeasure, and intrigue, evoked by the exhibit. Overall, about 40 percent of visitors' responses were emotional, with the tank of the African clawed frog generating the most frequent expressions of pleasure and a dead frog displayed to show internal organs showing the most frequent expressions of displeasure.

Conversation as a Tool to Understand Learning

This case study presents a snapshot of what people said while exploring the *Frogs* exhibit and how their conversations were categorized and explained. Allen notes that "hearing or reading visitors' complete conversations is a vivid experience that brings one right into the arena where real museum learning occurs. The transcripts are detailed, dense, and at times brutally honest, providing readers (be they developers, evaluators, or researchers) with a gritty sense of what engages and what doesn't. Personally, I found it a striking reminder of the power of choice in informal environments: visitors are choosing where to spend every second of their time, and exhibits that do not engage or sustain them are quickly left behind, however 'potentially educational' they may be" (Allen, p. 52).

Although Allen recognizes the power of this method, she also acknowledges how difficult it is to collect, transcribe, and interpret the data. She concludes, "Analyzing real-time visitor conversations in exhibitions is a fertile but costly complement to more traditional methods. Its strength is in bringing the researcher into the heart of the learning 'action' of the museum visit, and emphasizing learning as a process rather than merely an outcome" (Allen, p. 55).
(Adapted from Allen, S. (2002). Looking for learning in visitor talk: A methodological exploration. In Leinhardt, G., Crowley, K., and Knutson, K. (Eds.). Learning conversations in museums (pp. 259-303). Mahwah, NJ: Lawrence Erlbaum.)
[END OF CASE STUDY 4-1]

EXPLANATION: A LEARNING TOOL BETWEEN PARENTS AND CHILDREN

Allen provides insights into conversations that occur as pairs of any combination (two adults, a parent and a child, a grandparent and a child, for example) explore an exhibit. Other researchers have focused on how parents and children interact at a museum, with an emphasis on the role of explanation in enhancing the experience for the child. For example, developmental psychologists Maureen Callanan and Jennifer Jipson noted that parents often refer to prior experiences as a way to make an exhibit more relevant and meaningful. Overall, when parents mediate the exhibit for their children, the experience tends to be more beneficial.

A study by Kevin Crowley and his colleagues further illustrates the influence of parents on children in informal science learning environments. The researchers observed 91 families with children ranging in age from 4 to 8 years old as they explored an exhibit at the Children's Discovery Museum in San Jose, California. The exhibit focused on the zoetrope, a device that produces an illusion of action from a series of static images. The investigators found that children who engaged with their parents during their visit viewed the exhibit with more perceptive eyes. Their exploration was "longer, broader, and more focused on relevant comparisons" than that of children exploring the exhibit on their own. These results point to the key role that parents play in helping children select and identify appropriate details.

In addition, parents who have a background in science may be comfortable enough to use an exhibit as a starting point for sharing their knowledge. Below is an example of how a father, knowledgeable about simple machines, uses a Pulley Table to demonstrate for his son how this device works:

> *Well, mostly I was explaining to my son what it was doing. Showing that—for instance, there was one pulley that powered and the difference in putting the string on the smaller wheel as compared to the larger wheel, what it does to the other wheels. . . . Another boy walked up as well, and so I showed them the faster you turn it, the faster it plays, depending on the size of the pulley you use will also determine the power.*

Sometimes, however, parents may become too involved in the museum experience, which may in some ways limit children's access to cognitively complex tasks, as documented by researchers Mary Gleason and Leona Schauble. When 20 highly educated parents and their children (working in pairs) were asked to design and interpret a series of experimental trials to determine what factors cause a boat to be towed quickly through a canal system, the parents became immersed in the activity, looking up the results of previous trials and expressing their conclusions aloud. Although the parents did support and advance their children's reasoning, they tended to do the more challenging conceptual parts of activity, while delegating to the children the logistical components, such as releasing the boat into the canal and operating the stopwatch. As a result, it was the parents and not the children who made the greater gains in understanding the relationship between the boat and the canal system.

Other researchers have found that parent-child interaction not only can undermine children's engagement in the more challenging aspect of an activity, but also it can lead to misinterpretations of underlying phenomena. Parents can be aided in facilitating their children's learning by providing them with good lead-in questions, parent guides and similar types of resources, support, and guidance.

To examine more closely the dynamic between a parent and a child at a science museum, consider the following case study from research conducted by Kevin Crowley and Melanie Jacobs at the Pittsburgh Children's Museum. The case illustrates how parent-child teams interact and identifies the different types of rhetorical devices used.

[CASE STUDY 4-2]

A Conversation at the Museum

When parents take their children to museums, they generally try to make the experience as meaningful as possible for them. They point out key features, read the labels, and engage their child in a conversation about what they are learning. Often, too, parents take children to exhibits in subject areas in which they have expressed an interest.

Crowley and Jacobs's study, which comprised 28 families with children ages 4 to 12, focused on the nature of conversation between parents and children. This case study highlights one pair—a mother and her 4-year-old son.

In her first session with the researchers, the mother told them her family enjoys exploring science in a range of informal settings. The family had visited the natural history museum more than five times in the past year, as well as the local children's museum, the science center, and the zoo. The family also watches science-oriented television programs, reads books about science, and looks for science websites on the Internet.

The mom and her son began their museum visit at a fossil exhibit, which included a table of two sets of dinosaur fossils. One set includes authentic fossils, the second includes replicas. Each fossil had a card with information about the identity and age of the fossil and where it was discovered.

As the pair set out to explore the fossils, they were sitting in front of the fossil replicas. Laid out on the table were a dinosaur (Oviraptor) egg, footprint, tooth, claw, and coprolite (fossilized feces). The fossil that caught the boy's eye was the Oviraptor egg. The following conversation is a vivid illustration of a mom working hard to help her son enjoy an exhibit and learn from it.

Boy: This looks like this is an egg. *[He turns it over a few times in his hands.]*

Mom: OK, well this . . . *[picks up the card and glances at the label. She is using a "teachy" tone, which suggests that the boy is probably wrong and she is going to correct him and inform him what the object actually is.]*

Mom: That's exactly what it is! *[She appears surprised, speaking quickly in a more natural and rising tone of voice while turning to the child and patting him on the arm.]* How did you know?

Boy: Because it looks like it. *[He is smiling and appears pleased.]*

Mom: That's what it says, see look, *egg, egg* . . . *[pointing to the word "egg" on the card each time she says it and enunciating the way parents do when they are teaching children to read]* . . . Replica of a dinosaur *egg.* From the Oviraptor.

Mom: *[Turns gaze away from the card toward her child, putting her hand on his shoulder and dipping her head so that their faces are closer.]* Do you have a . . . You have an Oviraptor in your game! You know the egg game on your computer? *[Mom makes several gestures similar to the hunt-and-peck typing that a child might do on a computer keyboard.]* That's what it is, an Oviraptor.

Mom: *[Turns back to the card and points to the text on the card. She again starts speaking in her "teacher" voice.]* And that's from the Cretaceous period. *[pause]* And that was a really, really long time ago.

Throughout this conversation, the mother is going back and forth between "teaching" her child about the fossil and expressing pride that he is so knowledgeable. Her change in tone was particularly noticeable before and after she read the first label to him. She thought her son was wrong about the fossil—which accounted for the "teachy" tone before she read the card—but when she realized he was right, her voice switched to that of a proud mom. The researchers noticed the change in tone from formal to informal throughout this pair's visit to the museum.

The mother mediated the experience in several ways. First, to help her son place this experience in context, she reminded him about a dinosaur computer game that he played at home. Second, by "acting out" the way he types, she was giving a physical demonstration of the connection she was trying to establish to another family learning experience. Finally, she also changed some of the words on the card to make the explanation easier for a 4-year-old to understand. For example, the card had labeled the time frame as "Cretaceous Period, approximately 65 to 135 million years ago." The mother modified this explanation by simply saying that the Cretaceous Period was "a really, really long time ago."

The researchers noticed these same mediating strategies among other participants in the study. They also observed that other parents discussed each fossil's observable properties, its value and authenticity, and its anatomy. In addition, other parents made inferences about the size and function of the dinosaur based on the fossil they were examining.

At the end of each session, all the children involved in the study were asked to identify each of the nine fossils. To determine whether success in completing this task was associated with different levels of mediation, the researchers analyzed their findings in terms of how many fossils the children could identify based on their age and whether their parents provided high or low levels of mediation.

The results indicate that older children (ages 7-12) found the task of identifying fossils relatively easy. But even with this group, the more mediation they received, the better they did: children whose parents provided a low level of mediation identified 85 percent of the fossils correctly, and children whose parents provided a high level of mediation identified all the fossils correctly. A similar trend was found among younger children (ages 4-6).

Based on these data, it appears that higher levels of mediation, especially for younger children, result in more learning, defined as "the ability to identify fossils after going through the exhibit." What's more, further analysis of the data showed that, in particular, identifying what the object is and relating it to the child's past experiences had the greatest impact on learning.

This study makes a strong case for the value of parental guidance and mediation during informal science experiences. In particular, making connections to past experiences appears to solidify learning. Moving forward, professionals working in informal settings, including museums and out-of-school-time programs, can consider how to use these findings to strengthen the quality of their offerings.

(Adapted from Crowley, K., and Jacobs, M. (2002). Building islands of expertise in everyday family activity. In G. Leinhardt, K. Crowley, and K. Knutson (Eds.), Learning conversations in museums (pp. 333-356). Mahwah, NJ: Lawrence Erlbaum Associates)

[END OF CASE STUDY 4-2]

ROLES THAT SUPPORT LEARNING

Just as informal settings for learning vary tremendously, so do the practices in which facilitators, educators, and parents engage to support it. Even in everyday settings, facilitators can enhance learning. For example, a child's cause-seeking "why" questions are an expression of an everyday, intense curiosity about the world. Parents, older peers, facilitators, and teachers can and often do support these natural expressions of curiosity and sense-making, as revealed through studies about television watching discussed earlier. Evidence indicates that the more they do, the greater the possibility that children will learn in these moments. Recognizing expressions of curiosity and sense-making supports and encourages learning as productive and signals this value to learners (e.g., by listening to learners, helping them inquire into and answer their own questions, and involving them in regular activities that place them in contact with natural and designed phenomena and scientific concepts).

Roles that support learning can range from simple, discrete acts of assistance to long-term, sustained relationships, collaborations, and apprenticeships. For example, just by interacting with children in everyday routine activities (e.g., preparing dinner, gardening, watching television, making health decisions) parents, caretakers, and educators are often helping them learn about science. In addition, family and social group activities often involve learning and the application of science as part of daily routines. For example, agricultural communities regularly analyze environmental conditions and botanical issues. Even facilitators who are not experts in science (e.g., in after-school and community-based programs) can serve as intermediaries to informal science learning experiences. For example, the choice of pursuing a science badge in Girl Scouts may rest on the enthusiasm and assistance of a facilitating adult.

Productive science learning relationships frequently involve sustained individual inquiry but also intensive social interaction with interest groups and in mentoring relationships with experts, as revealed through Diane Miller's commitment to the teens in the program sponsored by the St. Louis Science Center in Chapter 3. In some cases, learners may develop a relationship with experts, who help them refine their science understanding and skill deliberately over sustained time periods. Seasoned science enthusiasts may serve as de facto mentors for newcomers in hobby groups (e.g., amateur astronomy, gardening). Distributed and varied expertise in groups allows less knowledgeable individuals to interact with more knowledgeable peers and mentors. Frequently, the roles of expert and novice shift back and forth over time, on the basis of specific aspects of the inquiry in question.

* * * *

Research shows that learning is a social process, heightened by conversation and engagement with other people. In designed settings such as museums, studies have illustrated how parents and other caregivers can mediate the experience for their children, making it more meaningful. With the knowledge that social interactions help facilitate learners, designers of informal science experiences can develop activities that encourage interactivity, discussion, and reflection.

Everyday experiences such as watching television, an intrinsically passive experience, can result in more learning when children engage with others in questioning, explaining, making predictions, and evaluating evidence.[11] Thus, in a variety of ways, including family social activities and conversation, children may begin to learn about topics that are relevant to science, even when learning science is not an explicit goal of the activity.[12]

In the next chapter, we focus more closely on the interest and motivations of learners, which they bring to each informal science experience. Understanding how these variables impact learning can lead educators to develop more compelling exhibits, activities, or programs.

Things to Try

To apply the ideas presented in this chapter to informal settings, consider the following:

- Consider if there is an informal way to "listen in" to the kinds of conversations people are having in your setting. If so, pay attention to what they focus on. Are they showing any of the kinds of learning discussed in this chapter?

- Have you noticed that one type of exhibit or experience seems to elicit more conversation than others? If so, is there a way to incorporate those features into other exhibits?
- A graphic representation from *Frog and Toad Are Friends* elicited a strong response from exhibit visitors. Are there ways to include artifacts from popular culture in your setting that would be recognizable to large numbers of people and could stimulate personally meaningful conversation?
- Are there tools in place to help parents and other caregivers mediate the experience for their children? For example, are the objects clearly labeled, with easy-to-read explanations? Are there guides for caregivers to help them deepen the experience of their children? Does the layout of the experience make it easy for adults to discuss the exhibit's ideas with their children? Is it possible to have staff people available to help parents engage in conversation with their children?
- Are experiences in your setting designed to be explored together? In programs, is interaction with other people an integral part of the experience? In exhibits, is there enough room for groups to explore? If an exhibit is designed for one user or visitor, can others observe and engage in other ways, for instance through conversations? In programs, is cooperation and collaboration made part of the experience itself? Does the experience encourage group reflection, conversation, joint problem-solving, and other forms of social interaction and cooperation? Are these experiences designed to elicit learning about the others in a group and thereby allow for strengthening interpersonal relationships?

For Further Reading

Allen, S. (2002). Looking for learning in visitor talk: A methodological exploration. In G. Leinhardt, K. Crowley, and K. Knutson (Eds.), *Learning conversations in museums* (pp. 259-303). Mahwah, NJ: Lawrence Erlbaum Associates.

Allen, S. (2002). Transcripts of the research from the *Frogs* exhibit. Unpublished.

Callanan, M.A., and Jipson, J.L. (2001). Explanatory conversation and young children's developing scientific literacy. In K. Crowley, C.D. Schumm, and T. Okada, *Designing for science: Implications from everyday, classroom, and professional settings.* Mahwah, NJ: Lawrence Erlbaum Associates.

Crowley, K., and Galco, J. (2001). Family conversations and the emergence of scientific literacy. In K. Crowley, C. Schunn, and T. Okada. (Eds.), *Designing for science: Implications from everyday, classroom, and professional science* (pp.393-413). Mahwah, NJ: Lawrence Erlbaum Associates.

Crowley, K., and Jacobs, M. (2002). Building islands of expertise in everyday family activity. In G. Leinhardt, K. Crowley, and K. Knutson (Eds.), *Learning conversations in museums* (pp. 333-356). Mahwah, NJ: Lawrence Erlbaum Associates.

Ellenbogen, K.M., Luke, J.J., and Dierking, L.D. (2007). Family learning research in museums:

Perspectives on a decade of research. In J.H. Falk, L.D. Dierking, and S. Foutz (Eds.). *In principle, in practice: Museums as learning institutions*. Lanham, MD: AltaMira Press.

Gleason, M.E., and Schauble, L. (2000). Parents' assistance of their children's scientific reasoning. *Cognition and Instruction, 17*(4), 343-378.

Haefner, M.J., and Wartella, E.A. (1987). Effects of sibling coviewing on children's interpretations of television programming. *Journal of Broadcasting and Electronic Media, 31*(2), 153-168.

Pedretti, E.G., MacDonald, R.G., Gitari, W., and McLaughlin, H. (2001). Visitor perspectives on the nature and practice of science: Challenging beliefs through A Question of Truth. *Canadian Journal of Science, Mathematics and Technology Education, 4*, 399-418.

Reiser, R.A., Tessmer, M.A., and Phelps, P.C. (1984). Adult-child interaction in children's learning from *Sesame Street*. *Educational Communications and Technology, 32*(4), 217-233.

Web Resources

Center for the Advancement of Informal Science Education (CAISE): http://caise.insci.org/

Exploratorium: http://www.exploratorium.edu

Informal Science: http://www.informalscience.org/

Pittsburgh Science of Learning Center: http://www.learnlab.org/

BOX 4-1
PROVOCATIVE TOPICS, PRODUCTIVE DIALOGUE

In some instances, informal science platforms bring up provocative topics, encouraging responses that are revealed through verbal or written discussions. For example, postings on websites and blogs as well as newspaper articles on such topics as genetic testing and stem cell research prod readers to think about, discuss, and write about their own views and positions. At the Science Café in Chapter 1, participants were encouraged to articulate their own opinions about the scientific evidence related to global warming.

While museum exhibits often do not give visitors an opportunity to think about science in this way, an exception is the Ontario Science Center's *A Question of Truth* exhibit. The purpose of the exhibit is to consider the cultural and political influences that affect scientific activity. The exhibit focused on three themes: frames of reference (e.g., sun-centered versus earth-centered), bias (concepts of race, eugenics, and intelligence testing), and science and community (interviews with diverse groups of scientists). After conducting interviews with visitors to this exhibit, Erminia Pedretti found that most of them thought the exhibit contributed to their understanding of science and society, "applauding the science center's effort to demystify and deconstruct the practice of science while providing a social cultural context."

For example, a visiting student commented, "The exhibit makes us think a lot about our beliefs and why we think in certain ways. . . . I didn't think that the gene that affects the color of your skin was so small and unimportant. Most people don't think of things like that." Another student challenged the view of science as being amoral: "We view science as often being separate from morals, and it's kind of negative because it allows them to do all sorts of things like altering human life, and it may not necessarily be beneficial to our society. . . . Some scientists are saying, should we actually be doing this?"

According to Pedretti, such comments indicate that exhibitions like these are encouraging visitors to reflect on the processes of science, politics, and personal beliefs and articulate their views. They achieve this goal by personalizing the subject matter, evoking emotion, and stimulating debate through the presentation of information from multiple perspectives.

5
Interest and Motivation: Steps Towards Building a Science Identity

The lights dimmed in the dome-shaped theatre at the Tech Museum in San Jose, California. Many museums and science centers have such theatres, which are especially designed for IMAX films. These unique films are large-format cinematic experiences that make viewers feel as though they are part of the action. This effect is achieved by displaying the film on the walls and ceiling, enveloping the viewer in the experience.

Viewers are waiting for *Coral Reef Adventure* to begin. The movie unfolds, showing the beauty of the coral reef and explaining why this delicate ecosystem is endangered. The audience is quiet, moved by the cinematic experience. For the moment, many people are ready to learn more about this environment and take action to protect it. But after the movie ends, the music dies down, and the lights come on, what do people take away from the experience? Do they have a greater understanding of the topic? Are they poised to become environmental activists?

To try to answer these questions, evaluators conducted interviews with 28 people (15 from the Tech Museum and 13 from the Science Museum of Minnesota) immediately after they watched *Coral Reef Adventure* (produced by MacGillivray Freeman Films) and three months later. Although this study sample was small, the results reveal the potential power of this medium.

Most of the 28 respondents reported that right after seeing the film, they talked about it with each other and recommended it to their friends and family. But perhaps what is even more interesting is that 3 months later, 23 of the 28 people interviewed noted that the film had a lasting effect.

> *I've definitely thought about how coral reefs are endangered. The film made a strong impression on my thinking about the ocean environment.*
>
> *It reinforced my concerns about the environment and conservation. Most people don't realize what's happening. They should see this film.*
>
> *We* [my wife and I] *did a couple of things. One, we were motivated to locate a site on the Internet about helping to preserve the marine environment. We also became members of the Monterey Bay Aquarium. As active participants in a study group at the aquarium we've learned more about why those trees are beneficial to the coral reef. We've also become members of The Tech* [Museum in San Jose]. *Thanks for making such a wonderful film."*

These results are corroborated from an evaluation of *Dolphins,* another IMAX film. According to researcher Barbara Flagg, three months after viewing this film, about 20 percent of the interviewed sample reported that they had taken action related to preserving the ocean environment. One respondent said that "we've joined a group that regularly goes down to the beaches to help clean them up."

Based on these findings, it appears that IMAX films spark interest in a topic and, in some cases, motivate viewers to learn more or to take action. These films are an example of what informal science venues can do to bring in crowds and generate excitement about science. *(Adapted from the Summative Evaluation of Coral Reef Adventure: An IMAX® Dome Film. Post-Viewing Telephone Interviews by Art Johnson, Edumetrics, and conversations with Barbara Flagg)*

THE ROLE OF INTEREST IN INFORMAL ENVIRONMENTS

Informal environments are often characterized by people's excitement, interest, and motivation to engage in activities that promote learning about the natural and physical world. Typically, participants have a choice or a role in determining what is learned, when it is learned, and even how it is learned.[1] These environments also are designed to be safe and to encourage exploration, supporting interactions with people and materials that arise from curiosity and are free of the performance demands that people often encounter in school.[2]

Interest, as described in Strand 1, includes the excitement, wonder, and surprise that learners may experience and the knowledge and values that make the experience relevant and meaningful. Recent research on the relationship between affect and learning shows that the emotions associated with interest are a major factor in thinking and learning. Not only do emotions help people learn, but also they help determine what is retained and how long it is remembered.[3] In addition, interest is an important filter for selecting and focusing on relevant information in a complex environment.[4] People pay attention to the things that interest them; hence, interest can drive what is learned.

This has been borne out in several studies focusing on conservation, which indicate that an individual's prior interest and involvement in conservation may serve as a better predictor of their responses and actions than typical demographic variables, such as age, gender, ethnicity, or education. Visitors with high interest in conservation stopped at more of the exhibits in a conservation-themed aquarium exhibition,[5] and zoo visitors' emotional responses to animals were more closely associated with emotional or personality variables[6] than demographic variables.

People with an interest in science are likely to be motivated learners in science; they are more likely to seek out challenge and difficulty, use effective learning strategies, and make use of feedback.[7] These outcomes help learners continue to develop interest, further engaging in activities that promote enjoyment and learning. People who come to informal environments with developed interests are likely to set goals, self-regulate, and exert effort easily in the domains of their interests, and these behaviors often come to be habits, supporting their ongoing engagement.[8]

Cultivating interest and motivation is a high priority for many informal science educators and has been explored and documented extensively in research, evaluations, and the accounts of practitioners. Many experiences are designed to capture and sustain participants' interest. There is evidence that the availability or existence of stimulating, attractive learning environments can generate the interest that leads to participation.[9] In fact, interactivity, which was discussed extensively in Chapter 3, may be useful in part because it generates and holds participants' interest.

There are many research-based frameworks for understanding interest and motivation and the role they play in the learning process. One such framework intended to enhance the quality of museum exhibits was developed by museum evaluator Deborah L. Perry. The model has six components:

- *Curiosity*—The visitor is surprised and intrigued.
- *Confidence*—The visitor has a sense of competence.
- *Challenge*—The visitor perceives that there is something to work toward.
- *Control*—The visitor has a sense of self-determination and control.
- *Play*—The visitor experiences sensory enjoyment and playfulness.
- *Communication*—The visitor engages in meaningful social interaction.

The following case study drawn from Perry's work with the Children's Museum of Indianapolis suggests how the model can be applied to exhibit design.

[CASE STUDY 5-1]

Questions to Ask: Building Exhibits Based on the Motivation Model

The Indianapolis Children's Museum set out to create an exhibit about color. The idea was to use colored lights to show how they overlap to make white light. Called *The Colored Connection: Making Colored Lights,* the plan was to shine red, blue, and green lights over a white tabletop so that overlapping circles of color were projected. Where the red and blue mix, the overlap forms magenta; where the blue and green mix, cyan is formed; and where red and green mix, they form yellow. In the center of the overlapping circles is a large white area, indicating that when all these colors combine, they result in white light. The exhibit also includes a computer with options for selecting colors, as well as three color-coded switches that can be turned on and off.

The exhibit designers wanted to know what they could do to improve on this preliminary design. Perry's model led to the following questions and decisions, showing one way to put research into practice.

How can the exhibit pique visitors' curiosity?

The first attention-getting strategy was the different-colored lights shining on the table. Once visitors were engaged, the exhibit kept their attention by using the computer to ask a question. The twist, however, was that the answer was embedded on another computer screen. The search for the answer proved to be an effective way to sustain visitors' interest.

How can the exhibit give visitors a sense of confidence?

Because research has illustrated that people are more likely to pursue activities when they feel they will be successful, this exhibit promoted confidence by writing the labels in a breezy, easy-to-read style. The result was that at least one person in a group understood the main ideas and were able to explain them to other members of the group. In particular, the labels were effective in educating parents about the exhibit, who could then convey that information to their children. Many parents commented that they felt successful when they were able to teach their children about the science behind the exhibit.

How can the exhibit challenge visitors?

Along with clear explanations, the exhibit also included numerous opportunities to stretch visitors' thinking. For example, one part of the exhibit asked visitors to experiment with making hand shadows. Another part encouraged visitors to make specific colors by turning the lights on and off. In these ways, visitors could choose to challenge themselves if they felt that they had mastered the main concepts presented in the exhibit.

How can the exhibit promote feelings of self-determination and control?

By definition, museums are "free choice" settings, but having too many choices in one exhibit can be overwhelming, detracting from both the enjoyment and the learning that takes place. The designers of *The Color Connection* experienced this problem during the first iteration of the exhibit, when they installed buttons instead of switches for turning the lights on and off. The designers found that people spent long periods pushing the buttons and creating light shows; by doing so, visitors were not learning anything and, ultimately, not really enjoying themselves either. When the buttons were replaced with switches, the visitors still had control over the lights, but they could manipulate them in a context of learning about white light and how it is formed. Because their activities led to a clear goal, visitors could enjoy what they were doing and feel gratified that they were gaining some information about science.

How can the exhibit promote feelings of sensory enjoyment and playfulness?

In this exhibit, visitors had opportunities to crawl through the lights, put different objects in the lights, make hand shadows, and then make up stories about their hand shadows. Activities such as these remind visitors how much fun they can have simply by experiencing science-related phenomena.

How can the exhibit stimulate meaningful social interaction?

By talking to members of their group, visitors often end up teaching concepts to each other. This kind of exchange not only provides opportunities for learning, but it also builds confidence, which helps keep motivation alive.

This model represents one approach to museum design. It continues to evolve, and some areas remain to be fully tested. Still, it offers a way to consider using research to plan exhibits that are more likely to draw in visitors, keep their attention, and encourage them to talk with each other to share knowledge and questions. Setting such goals and then determining whether they have been met not only allows museum educators to document learning, but also provides feedback for improving the quality of their offerings.

(Adapted from D.L. Perry, Designing exhibits that motivate. ASTC Newsletter, 1992)

[END OF CASE STUDY 5-1]

Although the focus of the exhibit described in the case was on how to capture and hold visitors' interest, in the process the designers also provided experiences that supported other strands of science learning. Thus, this example vividly illustrates their interconnectedness. By designing activities that piqued the visitors' interest, enhanced their sense of confidence, and provided them with engaging challenges, the designers also helped support conceptual understanding (Strand 2) and scientific reasoning (Strand 3).

Elements of design discussed in Chapters 3 and 4 also are evident in the case. The inclusion of labels promoted social interaction and the sharing of expertise. Interactivity played an important role in maintaining interest and supporting learning. At the same time, the designers found that including too many choices was overwhelming and detracted from learning—a finding similar to one found at the Exploratorium after studying the interactive features of the *Glowing Worms* exhibit (Chapter 3).

CULTIVATING AND SUSTAINING INTEREST

Thus far, we have focused on interest as the initial spark that hooks people and encourages them to explore an informal science experience. But interest can involve something more than just a visit to a museum or an hour at the IMAX theatre. Among informal science educators, there also is a desire to build sustained interest that will bring people back to learn more.

Many researchers have developed models for the development of long-term interest. Renninger and Hidi provide a useful framework that differentiates between shorter term interest and more sustained, engaged interest. Their four-phase model describes how interest emerges and changes as an individual becomes more engaged through repeated experiences related to a topic.

In the first phase, *situational interest*, excitement or interest is triggered by the situation. The participant's positive responses to a topic are typically sparked by environmental features that have personal relevance or capture attention because they are unexpected or unusual. In phase two, referred to as *maintained situational interest*, the participant has repeated positive experiences that are sustained by the meaningfulness of the tasks and personal involvement. In phase three, *emerging individual interest*, the person's interest starts to extend beyond the informal learning experience, which at this point is not always needed to stimulate interest or engagement with the topic. In the final phase, a *well-developed individual interest* becomes evident by the person's choice to continue their involvement by joining clubs, reading books, or participating in other activities on the topic. When an individual reaches this phase, he or she is highly motivated to look for more ways to learn about the subject.

The notion that interest can be deepened and sustained through repeated experiences is important to think about when designing informal learning experiences for science. Some settings or activities may not lend themselves to cultivating sustained interest as much as others do. Short visits to museums or lectures may trigger excitement about a topic, but these experiences do not offer enough exposure for well-developed individual interest to emerge. In order for this level of interest to develop, longer term engagement and multiple experiences are likely to be necessary and may be easier to integrate into some settings than others. After-school programs or citizen science experiences that last for weeks or months may promote sustained engagement more readily. Strategies for extending and connecting learning experiences across time and place are discussed in detail in Chapter 8.

To illustrate how sustained interest can evolve over time, consider the next case study, which describes a community garden project that was sponsored by York College of the City

University of New York. Over a period of nine months, urban African American and Latino teens and their mentors worked together to build a community garden. About 40 teens were involved in at least one aspect of the project, and a core group of 15 (12 boys and 3 girls) was responsible for implementing the project from start to finish. Their efforts illustrate a deepening of interest over the life of the project and the outcomes that are possible when people become truly engaged.

[CASE STUDY 5-2]

Science in the City: An Innovative Project with Teens

At a New York homeless shelter in the South Bronx, teacher and researcher Dana Fusco was offered a unique opportunity. She was asked to work with a group of young people between the ages of 12 and 16 to develop a science project. The teens had already faced many difficulties in their lives; when this project started, the participants were living in the shelter with their families.

When Fusco started working with the teens, her goal was for the project to emerge from the kids' own interests and frame of reference. She didn't want to impose her ideas on them. Rather, she wanted them to experience a process of reflection and discussion about issues that were of concern to them. From that point, she hoped that they would be able to reach a consensus about a group project. Ideally, the project would be personally meaningful, reflect what they have learned about science, and would benefit the community.

Fusco began by asking the teens what problems they were familiar with. Immediately, they started talking about teen pregnancy, AIDS, gangs, violence, drug and alcohol abuse, and racism—all experiences within their personal frame of reference. To express their views, the teens created a group collage, which they hung up at their meeting place.

After the teens had articulated their concerns, the group turned their attention to what they could do to address these problems. Fusco shared with them other projects that had been undertaken by urban youth. She mentioned awareness campaigns, community cleanup projects, mural painting, and gardening. Then Fusco mentioned that the teens had permission to use the lot across from the shelter as the site for their project.

At their next meeting, the participants investigated the lot. Although it was strewn with garbage, drug needles, and other debris, the kids immediately recognized its potential. One boy recalled that at one time, people had planted "stuff" in the back of the lot, but the plot had been burned. Other kids decided that their first step in moving forward with a plan should be measuring the lot. They began by coming up with makeshift strategies. One boy counted the number of steps it took to walk across the lot. Another group counted the number of concrete blocks lining the fence.

After their initial investigation of the space, they started discussing how it could best be used. The teens suggested a basketball court, an archery range, a playground, or a community garden. Before making a decision, the kids formed four teams to explore the space more thoroughly. One team measured the lot precisely, while another recorded evidence of living creatures and documented the nature of the nonliving debris. A third team took photographs, and a fourth made sketches of the lot.

The teams communicated with each other; for example, the recorders let the photographers and artists know where to go to find artifacts so that they could be captured on film. During this phase, the groups met often to review their data and determine which idea made the most sense for the lot. They narrowed the list down to the following: a playground, a garden, a clubhouse, a penny store, a jungle gym, a sandbox, and a stage.

As part of the decision-making process, the teens developed conceptual drawings to illustrate each of these ideas. Perhaps because they had seen evidence that there had once been a garden in that space or as a result of their discussions, the group decided that a community garden would be the centerpiece of the lot, with other structures surrounding and enhancing it.

The next phase of the project involved developing the expertise they needed to execute their plan. At this point, the teens realized that talking with community activists and experts in the field would help them build their knowledge base. This phase of the project proved to be very productive because the advice given led to new questions and new insights. For example, after talking with an environmental psychologist, the group realized that each element in the design of the urban garden has its own set of requirements. So if they built a stage, they would also have to build benches for seating; if they wanted to plant a garden, they would have to make sure that the plants had enough water and sunlight.

These realizations led to a new set of activities, including visiting other community gardens and a school with a composting facility, writing to organizations that might donate supplies or technical assistance, and bringing in gardening experts to discuss the garden's design and to help them test the nutrients in the soil. "During this part of the project, the kids really began doing science," says Fusco. "That, along with the community service piece, was very empowering."

For their part, the kids were somewhat surprised that they had actually been allowed to follow through with a project with demonstrable results. In fact, one teen remarked that, when the project started, he thought it was "going to be a project, like in school, you know, like a fake project." Another boy added to that sentiment by saying, "Yeah, I didn't think we were actually going to do it until you started talking about picking up the garbage and stuff." They came up with a name that reflected their enthusiasm for the project and their new status as scientists: Restoring Environments and Landscapes, or REAL.

"We formed our own 'gang' as an alternative to those in the streets," explains Fusco. "The group came together because of the friendships that formed."

Over the course of several months, the REAL team continued to gather more information about the site. During a slide show presentation of outdoor spaces, the group saw elements that they had not thought of, such as a storage shed, a path wide enough for wheelchairs to navigate, and signs. One member of the group felt so strongly about the need for a shed that he designed a storage space to fit under the stage. (This feature was later incorporated into a model the group made of the space.) One of the girls in the group made a sign for the model that said, "Help keep our REAL garden clean!"

Meanwhile, other teens were focusing on the garden. To determine where to place their flowerbeds, participants charted the position of sunlight throughout the day. Based on that data, they positioned the seedlings. As new information became available, the teens continued to modify their design. They added a birdbath and pond to attract wildlife to the garden, and they made sure that garbage cans and compost bins were part of the plan, as well as picnic tables, chess tables, and paths wide enough for wheelchairs. All of these features were incorporated into the model the teens made of their garden.

After months of research and planning, the group was ready to move into the implementation phase. They decided to hold a Community Day, during which they would share their model with their parents and businesses in the community. They alerted the neighborhood to the event by making flyers and posting them. More than 50 parents, staff, volunteers, neighbors, and children came to the group's community-wide event.

Volunteers from the neighborhood also got involved. Some helped clear out the garbage and sort out the recyclable materials. A professional carpenter worked with the kids to build a new fence. Other teens dug out the pond and planted seeds and seedlings with an expert gardener from Trinidad.

During the day, several teens walked around and video-interviewed attendees, asking them how they thought the garden would help the community. Below are samples of the responses they received.

After-school coordinator: "It's gonna give us [a] sense of responsibility because we're transforming something. We're making something out of nothing. We're gonna be extra proud because we did it."

Parent volunteer: "It's gonna turn out to be beautiful. It's gonna help the children take care of the neighborhood by seeing beauty."

Teen participant: "It will help the community by giving kids a place to come. Instead of being out in the street and doing things they shouldn't be doing, they can come in here and just relax and enjoy themselves."

Teen participant: "Because we need to [do] something for these kids right now. Things are not going good right now. Because you know how New York is filled with violence? So an event like this right here, it helps get away from all the violence."

By the time the project ended, the REAL team had planted tomatoes, peppers, and flowers in the garden and had built an arbor for the vines. "I really like doing community projects—it gives me a sense of responsibility and gives me a good feeling about helping people in the community," one participant remarked. A sense of ownership of the project from beginning to end combined with the opportunity to become involved in and contribute to the community resulted in a sense of accomplishment and deep commitment to the work. *(Adapted from "Creating Relevant Science through Urban Planning and Gardening, Dana Fusco, York College of the City University of New York, June 2001.)*

[END OF CASE STUDY 5-2]

Applying the Interest Development Model to the Community Garden Project

The community garden project is an excellent illustration of how interest can be cultivated and deepened over time. The project began with Fusco making the experience relevant for the urban teens, triggering their excitement (phase 1). When they realized that she was sincere and wanted them to take charge of the project, they began taking the work even more seriously, as reflected in the teen's comment that he didn't think they were really going to do anything until he saw evidence of action. At this point, the group was moving into phase 2, a sustained interest in the project and a growing curiosity about what steps they needed to take before they could start creating the garden.

Over the course of several months, the project took on a life of its own. The teens became more engrossed in the activity, taking the initiative to contact experts in the field, research plants, invite the local community to help with the project, and to roll up their sleeves to do the work to create their urban garden. Because the project continued over a relatively long period of time, the young people could cultivate their interest and even develop some expertise in gardening, construction, or fundraising. Over time, many of them became self-motivated, empowered by the fact that they were working in their own community and making a difference. At this point, their interest came from within and not from the environment.

It is important, too, to note that the project emerged from the members' own life experiences, which included concerns about gangs and violence. What happened over the course of the project is that the REAL team became a different kind of gang, as evidenced by such comments as "You down with REAL?" and "I'm getting REAL painted on the back of my denim jacket." This new "gang" developed its own mini "culture" of science within the larger community. The science gang now looked at the world differently and saw science as a way to reduce violence, create beauty, and bring disparate members of their neighborhood together.

Relating the Community Garden Project to the Strands

While the focus of this chapter has been on the motivational aspects of learning emphasized in Strand 1, this case also illustrates the interconnectedness of the strands. For example, in the teens' discussions about the lot and the viability of their ideas, they were demonstrating scientific reasoning skills (Strand 3). Further indications of the development of these skills emerged as they were determining what to plant based on the position of the sun at different times of the day and the quality of the soil.

As the teens prepared the lot for their garden, they used tools to track the living things occupying the lot and to determine the composition of the nonliving debris (Strand 4). Through the experience of using some of the basic tools of science, it can be argued that they were building a community grounded in the culture of science (Strand 5).

These tentative conclusions and their correlation to the strands are based on the observations of Dana Fusco, the project leader. Her views of the learning that occurred during this project reflect the sociocultural perspective on learning, which we summarized in Chapter 2 (Fusco, p. 872):

> The result [of the project] was not only the individual learning of science knowledge but the creation of science (and sciencelike) discourses, tools, and practices that had a real purpose within people's everyday lives. . . . What this suggests to me is that as youth, science, and community interact, the potential for change occurs at many levels—within the person, within the physical and social environment, and within the culture of science and science education. . . . Changes within the participants' ways of talking, thinking, and doing science occurred alongside practice and the creation of a science in which they would help minimize violence, beautify the community, and foster social and community gatherings and interactions.

WELL-DEVELOPED INTEREST AND CHANGES IN IDENTITY

As we saw in the interest development model, the last phase is "well-developed" individual interest, in which an individual chooses to engage in an extended pursuit in a particular area. When taken to its logical conclusion, the endpoint of this model is a change in identity on the part of the learner. For example, an individual who dabbles in gardening becomes so engaged by the activity that his or her identity becomes that of a "gardener." Such changes occurred in the teens who participated in the community garden as well as those who were part of the long-term program at the St. Louis Science Center (Chapter 3).

Identity, as described in Strand 6, includes the learners' sense that he or she can do science and be successful in science.[10] Identity is often equated with a subjective sense of belonging—to a community, in a setting, or in an activity related to science. The changes in community affiliation and related behaviors that can signal changes in identity usually require extended time frames of involvement with a program or community.[11] A sense of competence or belonging can be experienced retrospectively when reflecting on past events; it can be experienced in relation to current activities; and it can be projected into the future through imaginative acts regarding what one might become.

Identity can be viewed as both a critical factor in shaping educational experiences and a goal into which a broad range of learning experiences can feed. And it is an important element for all learners. While discussions of identity draw on widely recognized ethnic and cultural identities, promoting identification with science learning is an important issue for learners from all backgrounds.

Although researchers in the field generally agree that identity affects science participation and learning,[12] there are varied and disparate theoretical frameworks that address issues of identity. Some conceptions of identity emphasize personal beliefs and attitudes measured by the degree to which participants endorse such statements as "I have a good feeling toward science" or "I could be a good scientist."[13] Other conceptions of identity focus on the way that it is created through talk and other features of moment-to-moment interactions that position people among the roles and statuses available in particular situations.[14] This latter conception emphasizes that the type of person one can be in a setting—e.g., competent, skilled, creative or lacking in these qualities—depends on the way these types are defined in a social context; these identities are fluid and can change from setting to setting. The identities assigned to individuals in different settings are reinforced by the ways that people interact with material resources (e.g., instruments, tools, notebooks, media) and other participants (e.g., through speaking, gesture, reading, writing).[15]

There seems to be a strong relationship between science-related identity and the kinds of activities people engage in, usually with others. For example, parents who want to develop a particular family identity are able to quickly adapt the general museum experience, as well as specific content, to reinforce the desired identity. Everything from expectations ("We don't bang on the computer screen like that") to personal narrative history ("Do you remember the last time we saw one like that?") can be used to reinforce the values and identity of the family.[16]

But identity is not always the *result* of interest. John Falk and his colleagues from the Institute of Learning Innovation (ILI) have found that in some instances, it may also be the driving force, motivating people to join an informal science activity and shaping how they engage with it.

These ideas are based on a model of identity previously developed by Falk and his colleagues. This model suggests that visitors bring personal identities—as *explorers*, *facilitators*, *professionals and hobbyists*, *experience seekers*, or *spiritual pilgrims* (now referred to as rechargers)—to informal science settings. The researchers thought that these identities might be predictive of whether visitors experienced immediate and longer term changes in their attitudes and knowledge.

To test these ideas, Falk and his team embarked on a three-year collaboration with the Association of Zoos & Aquariums (AZA) and the Monterey Bay Aquarium. Specifically, they were interested in finding out if visitors to zoos and aquariums left with a greater appreciation of and deeper commitment to animal conservation.

Using a variety of research instruments, the team set out to test how many visitors could be categorized into the five identity categories and if each group showed distinctive behaviors during their visit. After collecting data from more than 5,500 visitors to 12 AZA-accredited zoos and aquariums, they found that although people had many reasons for visiting that did not fit neatly into a single category, the majority (55 percent) did have one dominant identity-related motivation that predicted how they experienced the setting and what they derived from it.

Explorers, who according to Faulk's model are curiosity-driven and interested in learning more as a result of the zoo or aquarium experience, were one of two groups with a dominant motivation. They were satisfied with the chance to see animals and learn more about them. They did not, however, report greater knowledge as a result of visiting or changes in attitudes about conservation, including their ability to make a difference through practicing it.

Facilitators, the second group with a dominant motivation, focused on helping the members of their social group enjoy the experience and learn from it. In this study, facilitators were looking for a social experience that benefited members of their social groups. Parents, for example, reported wanting to ensure that their children enjoyed their visit.

Professional/Hobbyists, though a small group of visitors (10 percent), are important because they feel connected to zoos and aquariums, largely because their offerings match the particular interests of this group. Professional/hobbyists reported looking for specialized programs, such as photo tours, dive trips, how-to workshops, and theme nights.

Experience Seekers enjoy new experiences and visit museums and other sites that are considered to be important. In this study, they made up 8 percent of the visitors and reported visiting as tourists or to support the community. They were the only ones to show a scientifically reliable positive gain in knowledge as well as a change in attitudes toward conservation. This finding could be explained by the fact that they arrived with the least amount of reported knowledge and the lowest expectation for their visit.

Rechargers are expected to be looking for contemplative and/or restorative experiences. In this study, 4 percent had such motivations, reporting that they wanted a place to think and get away from the noise and activity of the city. Overall, they reported visiting aquariums more than zoos.

Most visitors (61 percent) said that their zoo or aquarium experience supported and reinforced their values and attitudes toward conservation. But many (54 percent) said their visits prompted them to reconsider how they can affect environmental problems and support conservation. This shift indicates that they began to see themselves as part of the solution. Almost half (42 percent) of all visitors believed that zoos and aquariums play an important role in conservation education and animal care, and a majority (57 percent) of visitors said that their experience strengthened their connection to nature.

Identity as the Gateway to Deeper Engagement with Science

This research suggests that pinpointing identity-related motivations behind visits to zoos and aquariums could help educators figure out ways to better meet the needs of their visitors. For example, because explorers thrive on novelty, a way to reach them is to offer temporary exhibits or in-depth programs, as well as more challenging experiences. Opportunities for social interaction could be expanded for facilitators by offering meetings with staff and a designated place to go debrief after their experience. Similarly, experience seekers would enjoy a unique program that surpasses other local attractions, and professionals/hobbyists can be tapped to serve as volunteers. For rechargers, areas for reflection could be created and programs offered at quieter times of the day or year.

Although this model was tested in a museum setting, it also can be applied to other informal venues—even an everyday setting like an individual's home. The following case illustrates this point.

[CASE STUDY 5-3]

An Environmental Pioneer: Experimenting with a New Way of Life

Gabe Schwartzman is passionate about the environment. Growing up in suburban Maryland, where driving is a way of life, he has become increasingly concerned about the impact of car emissions on the environment. So he decided that he could make a difference by becoming self-sustaining in one area of his life. With some input from his cousin, Schwartzman set out to learn how to produce biodiesel fuel, which is made from oil left over after frying food.

"My cousin sent me a book, *From the Fryer to the Fuel Tank*, which explains how to make your own fuel," Schwartzman explains. After reading and rereading this book, as well as spending several months researching the subject and identifying a supplier (a local Chinese restaurant), Schwartzman began production.

In the basement of his parents' house, Schwartzman set up a makeshift lab. Wearing a leather apron, gloves, and goggles, he began the rather messy process of making biodiesel fuel. After separating the food remains left in the oil, he measured the pH. "It turned out to be a lot harder than I thought," Schwartzman admits. "Figuring out the correct pH for each container of oil took a lot of time."

Although the process was frustrating, it never occurred to him to give up. "I was determined to find out whether biodiesel fuel was an option for suburban drivers," says Schwartzman. He also was confident that he could succeed because he had worked on projects like this before. "He built a rickshaw when he was only 13," his mother remarks. "Once he sets his mind to something, there's no stopping him."

Schwartzman made his own fuel for a while and saved quite a bit of money; it cost only a dollar to fill up the tank of his 1980 Volvo. But after making the fuel for several months, Schwartzman abandoned the project. "My car died, but I also decided that biodiesel fuel wasn't the best option for drivers in busy metropolitan areas," he explains. "The process is time-consuming, and the fuel needs are too great. But I wouldn't have known that if I hadn't seen the project through to the end. I'm glad I did it. I learned a lot."
[END OF CASE STUDY 5-3]

Schwartzman is an example of a highly motivated and engaged learner, one whose interest far exceeds that of most people. Not only did he learn about a new topic—how to make biodiesel fuel—but he also produced the fuel. And in the face of many obstacles, he persisted, showing a strong commitment to learning.

This story also illustrates how Falk's model can be applied to a nonmuseum setting. Schwartzman is both an explorer—constantly on the lookout for new ideas and experiences—and a professional/hobbyist—an amateur scientist who has attained a high level of expertise.

* * * *

This chapter discussed motivation and interest (Strand 1) in greater depth. In particular, two models were discussed that can be applied to designing an interesting and motivating informal science experience. Perry's six-component motivation model describes factors to consider when designing effective museum exhibits. Renninger and Hidi's interest development model describes how the environment may initially need to spark interest before personal motivation develops. A community urban garden project for teens was used to illustrate the four phases of interest development, as well as other kinds of learning described by the strands of learning.

Interest and identity are intertwined. The chapter concludes with a study describing the role of identity in motivating behavior, with suggestions of how informal science settings can make use of this information. The identity model also is flexible enough to be used in evaluating learners' behavior in everyday settings, as illustrated by Gabe Schwartzman's experience making biodiesel fuel.

Throughout the book, we have seen how learners do not gain new knowledge and insights in a vacuum. Rather, their learning is enhanced through engagement with others, experimentation, and interaction with museum artifacts. The next chapter reinforces this point by exploring how an individual's culture affects the way he or she approaches informal science learning environments. This diversity of perspectives needs to be recognized when designing these learning experiences.

Things to Try

To apply the ideas presented in this chapter to informal settings, consider the following:

- *Share Perry's six-component motivation model with staff and consider whether or how it can be used in your setting.* The case study of exhibit design can be used as a starting point for discussing whether Perry's motivation model would be helpful in your setting. If so, choose a particular exhibit or program and discuss how each component of the model could be used to design or modify a participant's experience in your setting.
- *Consider ways to excite interest in a program or exhibit.* The research indicates that piquing participants' interest is an important step in bringing about self-motivation. Discuss how this finding applies to your setting. What strategies have you tried in the past? Based on what you have read, are there other strategies that might be effective in getting participants' attention, sparking their interest, and sustaining it?
- *Assess experiences for unintended negative emotions.* Just as positive emotions can trigger learning, negative emotions can be a turnoff. Which aspects of the informal science experience that you offer might be confusing, overwhelming, or inadvertently

unpleasant in any other way? Have you layered the experience sufficiently to make it beneficial and enjoyable to visitors or audiences who approach it with different levels of interest?

- *Consider how your experience can tap into the multiple identities that visitors bring to the experience.* Falk's model explains five identity-related motivations that lead visitors to different kinds of experiences. Can this model be applied in your setting to tailor experiences that meet the needs of these different kinds of visitors? Discuss how this could be accomplished; one possibility is to design an exhibit or experience specifically for one of these types of visitors. Assess whether it was successful and how additional experiences could be designed for other types of visitors highlighted in this model. Consider the overall balance of experiences your institution provides for the five situated identities? Are there opportunities for reflection and restoration in a noisy and active environment? Are there opportunities to explore and seek new experiences even for those who visit multiple times? Can you provide visitors with guidance that allows them to choose experiences in your setting that best align with their situated identity (a parent guide for young children; a guide for a one-time visitor on "what not to miss," etc.)?

For Further Reading

Fusco, D. (2001). Creating relevant science through urban planning and gardening. *Journal of Research in Science Teaching. 38 (8),* 800-877.

Falk, J.H., Reinhard, E.M., Vernon, C.L., Bronnenkant, K., and Heimlich, J.E. (2007). *Why zoos & aquariums matter: Assessing the impact of a visit to a zoo or aquarium.* Silver Spring, MD: Association of Zoos & Aquariums.

Hidi, S., and Renninger, K.A. (2006). The four-phase model of interest development. *Educational Psychologist*, *41*(2), 111-127.

Johnson, A. (2003). Summative Evaluation of Coral Reef Adventure—*An IMAX® dome film.* Post-Viewing Telephone Interviews. Unpublished report.

Perry, D.L. (1994). Designing exhibits that motivate. In R.J. Hannapel (Ed.), *What research says about learning in science museums* (Vol. 2, pp. 25-29). Washington DC: Association of Science-Technology Centers.

Renninger, K.S. (2007). *Interest and motivation in informal science learning.* Unpublished report.

Web Resources

Cool Fuel: Brew It Yourself http://www.washingtonpost.com/wp-dyn/content/article/2008/06/30/AR2008063002280.html

Designing Exhibits That Motivate
http://www.selindaresearch.com/Perry1992DesigningExhibitsThatMotivate.pdf

Creating Relevant Science through Urban Planning and Gardening
http://www3.interscience.wiley.com/journal/85513199/abstract

6
Assessing Learning Outcomes

A key to designing informal experiences to support learning is to clearly articulate the goals for a particular experience. That is, what should participants take away? In this book we define science learning in terms of six strands, which encompass excitement and interest spurred by an aspect of an informal experience, understanding scientific knowledge, engaging in scientific reasoning, reflecting on the nature of science, increased comfort with the tools and practices of the scientific community, and identifying with the scientific enterprise. In any given experience, particularly those that are very brief, it may be impossible to touch on all six strands. However, the design process should include explicit decisions about which outcomes are of primary interest.

Although informal science settings do not use the same tools to assess learning as schools do—tests, grades, and class rankings, for example— researchers, evaluators, and practitioners are nonetheless very interested in assessing how informal experiences contribute to the development of scientific knowledge and capabilities. The nature of informal settings presents a unique set of challenges in this effort and the field struggles with theoretical, technical, and practical aspects of measuring learning. This chapter explores some of these challenges and the ways they have been addressed.

CHALLENGES OF ASSESSING SCIENCE LEARNING IN INFORMAL SETTINGS

The characteristics of informal learning environments make it very difficult to develop practical, evidence-centered ways to assess learning outcomes. For example, during a short trip to a museum, not only is assessment logistically complex, but the data gathered are hard to interpret. It can be difficult to separate the effects of a single visit from other factors that could be contributing to positive learning outcomes. And arranging for tests before and after the experience or setting up other traditional measures in many museums and science centers can be disruptive.

Another feature of informal science learning environments that creates challenges for assessment is that experiences are not prescribed or predetermined. Rather, the environments are learner-centered and much of what happens emerges during the course of activities. Because each visitor seeks out his or her own unique experience, it is extremely difficult to establish a uniform intervention or activity in order to assess it's impact on learning. Part of the problem, too, is the importance of not interfering with a unique, self-directed experience, because it is often what inspires learning in the first place. The challenge thus becomes how to document the learning that occurs while not sacrificing the freedom and spontaneity that is integral to the experience.

The collaborative and social aspects inherent in many informal experiences also pose a challenge for assessing learning. Participants in summer camps, science centers, family activities, hobby groups, and such are generally encouraged to take full advantage of the social resources available in the setting to achieve their learning goals. The team designing a submersible in camp or a playgroup engineering a backyard fort can be thought of as having implicit permission to draw on the skills, knowledge, and strengths of those present as well as any additional resources available to get their goals accomplished. "Doing well" in informal settings often means acting in concert with others. Thus, assessments that focus on an individual's performance alone may "undermeasure" learning because participants re not be able to draw on material and human resources in their environment, even though making use of such resources is a hallmark of competent, adaptive behavior[1]. In addition, assessing whether participants working in a group have grasped the science is important, but equally important is measuring the role that collaboration and problem solving have played in learning. Teasing out this variable from individual assessment has proven to be difficult.

Despite the difficulties of assessing outcomes, researchers have managed to do important and valuable work. Many of these approaches rely on qualitative interpretations of evidence, in part because researchers are still in the stages of exploring features of the phenomena rather than quantitatively testing hypotheses.[2] Yet, as a body of work, assessment of learning in informal settings draws on the full breadth of educational and social scientific methods, using questionnaires, structured and semistructured interviews, focus groups, participant observation, journaling, think-aloud techniques, visual documentation, and video and audio recordings to gather data.

A Holistic Approach to Assessment

Although many people may think that assessment is a review completed at the end of a project, it is actually much more than that. Assessment is a process that begins at the outset of a project and continues until the end. It encompasses determining whom the project is for, what learning goals are reasonable for that audience, and what instruments are going to be used to determine whether those goals have been met. Randi Korn, an evaluator in the informal science community, expresses this idea as follows: "There is an invisible link between planning and evaluation. Begin planning with the end in mind."

As stated above, the first step in the assessment process is identifying the learning goals, followed closely by determining the audience. But this determination also is complex. It is not sufficient to simply target a demographic group, such as teenagers or Latinos, as the audience. It is equally important to try to understand what knowledge, skills, and beliefs such a group brings to the learning situation. For this reason, key stakeholders, including representatives from the institution or organization and members of the community, must be brought into the planning process. This phase of program development is called *front-end evaluation.*

There are many examples of front-end evaluation in this book. In Chapter 5, for example, we outlined questions that are part of Deborah's Perry's motivation model. By discussing these questions with focus groups before finalizing an exhibit, designers can help ensure that it is well suited to the audience. In the next chapter, we explore how individuals planning programs for different cultural communities recruited representatives from those communities to serve on advisory committees charged with planning specific events. A good first step in the planning process is getting to know the audience. Working with stakeholders representing a group or getting feedback from those viewing an exhibit are both ways of accomplishing this goal.

During the design and development phase of a project, a *formative evaluation* often is conducted. The purpose of this step is to determine what is working—or not working--before the program is completed. In *The Mind* exhibit developed at the Exploratorium, for example, Erik Thogersen went through a formative evaluation phase by prototyping different possibilities for the exhibit with visitors to the museum. Some ideas became part of the finished exhibit, and others were rejected. This process provides a midpoint check for developers so that they continue to question their assumptions about the project, consider whether goals and objectives are being met, and make necessary changes before the project is completed.

The final phase of the assessment process is the *summative evaluation.* Conducted after the project is completed, the purpose of this evaluation is to document whether the learning goals established at the beginning of the project were met and where there is room for improvement. During this phase, some unplanned-for learning outcomes also are noted.

Factors Involved in Data Collection

As with front-end evaluation, summative evaluation reports also have been referred to in many of the book's case studies. They have been used to document the learning that occurred. In those discussions, we usually mention the types of assessment instruments used to collect data.

Selecting the most appropriate instrument to measure specified learning goals is another challenge of the assessment process. In making this determination, designers need to take into consideration three variables. First, the assessment instrument must be designed to measure the specified learning goals. So, for example, if the exhibit or program aims to help participants become excited about science and learn science content, then the assessment instrument must have mechanisms in place to measure both of these goals.

Second, assessments should fit with the kind of participant experiences that make informal learning environments attractive and engaging. Any assessment activities undertaken in these settings should not undermine the very features that make for effective learning. Third, the assessments must be valid; that is, they should measure what they purport to measure and align with opportunities for learning that are present in the environment.

Keeping these variables in mind, let's consider how information was collected in some of the examples included in this book. WolfQuest, the computer game discussed in Chapter 1, used online surveys to collect data about the project. The survey asked questions related to the learning goals, as the first variable suggests. Considering this was an online experience, it is fitting that an online assessment instrument was used. Because of the nature of the activity, assessment did not interfere with the experience. Finally, the assessment was valid; questions asked on the survey align with the project's goals.

Through the WolfQuest online survey, evaluators asked participants what they knew about wolves before playing the game and what they learned as a result of the game. This pre-post strategy is often used to document changes in learning as a result of the informal experience. When evaluators are interested in finding out whether information was retained over time, they typically get in touch with visitors through phone or e-mail three or four months after the experience to see what they remember and whether the experience had a lasting impact. The evaluation conducted for the IMAX film *Coral Reef Adventure* (Chapter 5) used this strategy and collected compelling information about how the film led to changes in attitudes and behaviors.

Another, more complex form of data collection is taping visitors' conversations. As discussed in Chapter 4, this approach is logistically difficult to execute, and interpreting the data is equally challenging. Nonetheless, the information collected can be rich and revealing, indicating what visitors are thinking and feeling and whether the experience has evoked powerful memories. Aware that gathering data could interfere with the learning experience, the researchers investigating conversations in the frog exhibit at the Exploratorium and those "listening in" on conversations between parents and children thought long and hard about how to accomplish this so that visitors would have an authentic experience.

Precisely because of these challenges, collecting conversations is done less frequently than more traditional tracking and timing methods used to measure level of engagement with an informal science experience. Behaviors measured through these methods are labeled *stop* (planting the feet and attending to an exhibit for at least 2-3 seconds), *sweep rate* (the speed with which visitors move through a region of exhibits), and *percentage of diligent visitors* (the percentage of visitors who stop at more than half of the elements). The percentage also suggests benchmarks of success for various types of exhibit formats, such as dioramas or interactives. In *Cell Lab* (Chapter 3), evaluators noted that visitors spent considerably more time than usual at the wet-lab benches. This finding illustrated that when learning is more complex, more time is needed—even if fewer people can go through the exhibit in one day.

[SIDEBAR 6-1]

The NSF Evaluation Framework

Recognizing the challenges inherent in conducting valid assessments in informal settings, the National Science Foundation (NSF) has developed a flexible framework that can be used as both a planning and an assessment tool. NSF has identified five impacts, which are areas that can be identified and assessed in these settings. While the impact categories are similar to the strands, they offer a slightly different perspective. The impact categories are as follows:

- *Knowledge.* Similar to Strand 2 (understanding scientific content and knowledge), this impact refers to knowledge, awareness, or understanding that visitors can express in words or pictures that illustrate what has been learned during, immediately after, or long after a given experience.
- *Engagement.* Similar to Strand 1 (sparking interest and excitement), this impact focuses on the emotions evoked by the experience. These emotions can range from excitement and delight to negative feelings, such as anger or sadness.
- *Attitude.* Similar to Strands 1 and 6 (sparking interest and excitement and identifying with the scientific enterprise), this impact refers to a change in world view or an increase in empathy as a result of an experience in an informal setting.
- *Behavior.* This impact refers to projects whose purpose is to change visitors' behaviors over the long term. Often these changes are sought after experiencing environmental or conservation projects.
- *Skills.* Similar to Strand 3 (engaging in scientific reasoning) and Strand 5 (using the tools and language of science), this impact focuses on the skills of scientific inquiry, such as observing, asking questions, predicting, testing predictions through experimentation, collecting data, and interpreting them.

The framework also includes a suggested chart that can be used to track impact, the category in which it falls, objectives, and evidence (or not) of the impact. Program designers and educators can identify as many impacts, in as many of the major categories, as they believe fit their project. The advantage of this system is that it provides an easy way both to plan a project and then to document its overall effect. It is a useful formative evaluation tool.

[END SIDEBAR 6-1]

Assessment Still a Challenge

Although the steps in the assessment process provide a useful guide to evaluators, it is still challenging to conduct assessments in informal settings. To continue to improve the approaches in place, researchers from the Institute for Learning Innovation suggest taking into account the following core assumptions about the nature of informal science learning outcomes. A rough consensus about these ideas is emerging in the informal science community.

o *Outcomes can include a broad range of behaviors.* In addition to the key types of individual outcomes currently being investigated, research could be designed to allow for varied personal learning trajectories and outcomes that are complex and holistic, rather than only those that are narrowly defined.

o *Outcomes can be unanticipated.* Outcomes can be based on the goals and objectives of the program, or they can be unplanned and unanticipated, based on what individual learners find to be most valuable. Research can target outcomes that emerge from learners' experiences, not only those that are defined in advance.

o *Outcomes can become evident at different points in time.* While short-term outcomes have long been used to assess the impact of informal learning experiences, it is becoming increasingly evident that these experiences can have enduring, long-term impacts as well.

o *Outcomes can occur at different scales.* To date, most outcome measures are focused on determining how the individual was influenced by the experience. But it is also useful to consider how the entire social group was influenced. For example, did group members learn about one another? Did they reinforce group identity and history? Did they develop new strategies for collaborating together? In addition, outcomes can be defined on a community scale, measuring how an activity, exhibition, or program affected the local community.

In practical terms, the kinds of assessments that work best in informal settings are likely to be the ones that most closely match the setting's learning activities. Before drawing conclusions about whether a particular experience has led to a particular outcome, researchers and practitioners should ask themselves:

- Are the assessment activities similar in relevant ways to the learning activities? At the Cell Lab stations, the process of doing the activity also served as an assessment of how well the participants understood the point of the experiment and how to interpret their results. Designing activities that can also serve as an assessment tool works particularly well in informal settings.
- Are the assessments based on the same social norms as those that promote engagement in the learning activities? For example, in assessing WolfQuest, the researchers used an online forum for assessment, which matched the nature of the activity.
- Is it clear that the learners have had ample opportunity to both learn and demonstrate desired outcomes? The teens working with young children at the homeless shelter in St. Louis (Chapter 3) illustrated how, over time, they not only learned relevant science content but also could demonstrate their new learning in multiple ways. In Chapter 8, which explores science learning across the life span, the time that teens,

adults, and seniors had to spend on their projects resulted in attainment of new knowledge and skills, as well as increased interest in science.

* * * *

This chapter discusses an approach that can be used as a guide in planning and executing assessments in informal science environments. A key element of this approach is the importance of up-front planning, during which it is essential to set goals for the project and get to know the audience. Knowing what the learning goals are at the beginning of the project is crucial to its success.

The next phase of the assessment process is formative evaluation, which provides some data about the project before it has been completed so that changes can be made. The final phase, summative evaluation, is a final report that specifies whether the learning goals were met.

The chapter also describes several instruments that are frequently used to collect assessment data, drawing on the book's case studies for examples of these instruments . Ideally, assessment instruments are selected because they work in the setting and do not interfere with the learning taking place.

Although much progress has been made in understanding how to conduct assessments and interpret data gleaned from the process, more work needs to be done. By continuing to build consensus about the best way to perform assessments, it will become easier to develop effective ways to document the learning that occurs in a wide range of informal science settings.

Things to Try

To apply the ideas presented in this chapter to informal settings, consider the following:

- *Discuss the role of assessment in your setting.* This chapter has emphasized the increasingly important role that assessment plays in informal science settings. How do these ideas apply to your setting? Consider how these ideas could help improve the assessments currently being conducted in your institution.
- *Is there a clear link between planning and assessment?* Evaluators are realizing the importance of connecting the planning process to evaluation goals. Has this idea gained traction in your setting? Do you see ways to link the two processes? Do you align your resources with your goals?
- *Think about how to refine assessment instruments.* This chapter discussed many assessment instruments and how they fit the outcomes they are designed to measure. Consider whether assessment instruments currently being used at your institution can be modified and improved on based on these ideas. Consider also whether you have defined appropriate goals, outcomes, and indicators that guide assessment?
- *Consider unintended outcomes.* Assessment can focus on clearly defined outcomes, but in informal settings there are often far more outcomes as seen by your participants than can be programmed for or can be assessed. Do you have a full understanding of the learning benefits that your audiences or participants derive? Talk to your participants or your audiences about the way they see themselves benefiting or being better off.
- *Create a culture of curiosity.* Getting it right is better than learning from assessment at the end about things that could have been done better. A culture of audience engagement in the process of planning, designing and implementing visitor

experiences is important to achieve success. Create a culture in which participant voice and audience input are a source of information and inspiration and where asking visitors can be done on the fly.

- *Think about the "silent majority".* When talking to participants and audiences, though, make sure you hear all voices. It is easy to focus on just those who provide lavish praise or harsh criticism, and satisfying either one of these two extremes may not be beneficial for many others.
- *Share with others.* There are many ways today to share your assessment experience with others. Make your insights available to the community of informal science educators and draw from other experiences as well.
- *Utilize external resources.* A variety of online resources are now available that support assessment and evaluation, and articles and books that address these issues are of increasing quality. Websites that archive resources in informal science learning and teaching (www.informalscience.org), afterschool program assessment (http://atis.pearweb.org/), or visitor studies (www.visitor studies.org) provide gateways into the assessment and evaluation community.

For Further Reading

Clipman, J.M. (2005). *Development of the museum affect scale and visit inspiration checklist.* Paper presented at the annual meeting of the Visitor Studies Association, Philadelphia. Available at: http://www.visitorstudiesarchives.org

Falk, J.H., Reinhard, E.M., Vernon, C.L., Bronnenkant, K., Deans, N.L., and Heimlich, J.E. (2007). *Why zoos and aquariums matter: Assessing the impact of a visit.* Silver Spring, MD: Association of Zoos and Aquariums.

Garibay, C. (2005, July). *Visitor studies and underrepresented audiences.* Paper presented at the annual meeting of the Visitor Studies Association, Philadelphia.

Institute for Learning Innovation (2007). *Evaluation of learning in informal learning environments.* Paper prepared for the Committee on Learning Science in Informal Environments.

National Research Council (2009). Introduction. Chapter 3 in Committee on Learning Science in Informal Environments, *Learning science in informal environments: People, places, and pursuits.* P. Bell, B. Lewenstein, A.W. Shouse, and M.A. Feder (Eds.). Center for Education, Division of Behavioral Sciences and Social Science and Education. Washington, DC: The National Academies Press.

Web Resources

Center for the Advancement of Informal Science Education (CAISE): http://caise.insci.org/

Informal Science: http://www.informalscience.org/

Institute for Learning Innovation: http://www.ilinet.org

National Science Foundation Evaluation Framework:
http://insci.org/resources/Eval_Framework.pdf

Part III
Reaching Across Communities, Time, and Space

7
Culture, Diversity, and Equity

Native Waters: Sharing the Source is an exhibit at the Science Museum of Minnesota with a double message. Its goal is to share cultural views about water held by the tribal peoples of the Missouri River Basin as well as scientific concepts about the Missouri River and its watershed. The exhibit accomplishes its double-pronged goal through its design, informative text panels, and interactive features.

The exhibit is set up like an Indian tipi, with the inside space designated as a place to hear stories about native culture. A sculpture of a spring takes center stage, with four banners, pointing in the four cardinal directions (north, south, east, and west), emanating from the spring. Each banner is illustrated with native drawings and includes quotes from Missouri Basin elders and tribal members. Visitors learn about sunrise and sunset on the east and west banners and about the phases of the moon, which cut across geographical boundaries. On the tipi wall is the story of the Missouri River. It begins in the Rocky Mountains and travels east until it reaches Cohokia, a native settlement that once had a population of 50,000.

The story of the river is told as a blend of scientific and native elements. As the river moves eastward and downhill, seasonal changes affect its size, creating what often is referred to as its pulse. The clockwise movement of the river is balanced by counterclockwise movements brought on by seasonal changes. According to native lore, it also represents the idea that while traveling forward, one should also remember the past.

This example illustrates one strategy for closing the gaps that can exist between the beliefs, values, and practices of some communities and those embodied in Western science. By incorporating elements of native culture into a science exhibit, the designers are blurring the border between Western and native approaches to understanding the natural world, requiring all visitors to examine their own world views in one way or another.

An important value of informal environments for science learning is being accessible to all people. However, social, economic, cultural, ethnic, historical, and systemic factors all influence the types of access and opportunities these environments provide to learners.[1] Learning to participate in science—that is, developing the necessary knowledge and skills, as well as adopting the norms and practices associated with doing science—is difficult for many people. It can be especially challenging for members of traditionally underrepresented (or nondominant) groups.

The challenges of engaging nondominant groups in the sciences are reflected in studies showing that (1) inadequate science instruction exists in most elementary schools, especially those serving children from low-income and rural areas; (2) girls often do not identify strongly with science or science careers; (3) students from nondominant groups perform lower on standardized measures of science achievement than their peers; (4) although the number of individuals with disabilities pursuing postsecondary education has increased, few pursue academic careers in science or engineering; and (5) learning science can be especially challenging for all learners because of the specialized language involved.[2] Addressing these challenges requires rethinking what it means to provide “access” to science.

RETHINKING EQUITY

Striving for equity in science education has often resulted in attempts to provide better access to opportunities already available to dominant groups, without consideration of the cultural or contextual issues that must be taken into account. Science instruction and learning experiences in informal environments often privilege the science-related practices of middle-class whites and may fail to recognize the science-related practices associated with individuals from other groups. In informal settings for learning science, such as museums, some initiatives are aimed at introducing new audiences to existing science content by offering reduced-cost admission or bringing existing science programming that is already offered to mainstream groups to nondominant communities.

The logic of this view is that individuals from particular groups or communities have simply not had sufficient access to science learning experiences. To remedy that situation, educators deliver to nondominant groups the same kinds of learning experiences that have served dominant groups. However, simply exposing individuals to the same learning environments may not result in equity, because the environments themselves are designed using the lens of the dominant culture. For example, the signs and labeling of an exhibit may be in English only, or a program for families may be designed to accommodate the 1- or 2-parent family structure typical of many middle-class, white families, rather than the multigenerational, extended family structures that may be prevalent among other groups.

To achieve equity, practitioners must consider ways to connect the home and community cultures of diverse groups to the culture of science. Angela Calabrese Barton, professor of science education at Michigan State University, argues for allowing connections between learners' life worlds and science to be made more easily and "providing space for multiple voices to be heard and explored."[3]

An important first step toward designing more inclusive and genuinely equitable learning experiences in science is for educators and designers to recognize that they may be acting under assumptions that reflect the dominant culture of middle-class whites. As a result, the programs, activities, and exhibits they design may have narrow appeal and lead people from nondominant cultures to perceive them as directed by and designed for the dominant group. Cecilia Garibay, principal of the Garibay Group, points to a number of indicators identified through research that can support this perception, including the lack of diverse staff, a feeling that the content is not culturally relevant, and the unavailability of bilingual or multilingual resources. In fact, recent research with various cultural groups suggests that these issues result in nondominant communities feeling unwelcome in museums.[4]

Approaching these problems with outreach efforts may inadvertently reinforce the image of informal settings as being part of the dominant culture. The term *outreach* itself implies that some communities may be external to the institution. Promoting collaboration, partnership, and diversity in power and ownership may provide greater opportunities for nondominant groups to see their own ways of thinking and meaning-making—or making sense of what they are seeing and experiencing—reflected in informal settings.

To this end, making an apparently simple adjustment—such as translating labels into multiple languages—has been shown to make a significant difference. Not only does this practice help members of other cultures identify key elements in an informal experience, but it also facilitates conversation and enhances learning among intergenerational groups.[5] That said, it is important to point out that providing signage and labels in multiple languages can actually be a big undertaking for a museum. It requires adjustments to the exhibit development process and incurs costs for translation, proofing, and production. However, the additional investment is an important step toward providing more equitable learning experiences. Alternatively, another way to accomplish the same goal is to have a non-English-speaking mediator available to "talk through" the experience with the visitors.

Attention to language differences is only one component of designing for equity. It also is important to consider variation in beliefs, values, and norms of social interaction, such as variability in family structure, gender roles, and patterns of discourse (e.g., the role of questioning in a conversation). Many informal institutions nationwide are addressing these issues and modifying exhibits to reflect these differences. The next case study is one such example. It describes how a large children's museum, Children's Discovery Museum in San Jose, California, established an ongoing relationship with the city's growing Vietnamese population; through this partnership, the museum was able to develop a significantly more inclusive learning experience. The museum's work in this area highlights both its challenges and rewards.

[CASE STUDY 7-1]

The Vietnamese Audience Development Initiative

In 2002, the Children's Discovery Museum (CDM) launched its Vietnamese Audience Development Initiative to better understand San Jose's growing Vietnamese community and to develop strategies for helping the museum better meet the community's needs. San Jose is home to more residents of Vietnamese descent than any other city outside Saigon. After gaining experience working with another cultural group—the Latino community—museum staff decided to begin working with the Vietnamese community. They also recognized that the Vietnamese community represents a fairly low percentage of its visitors and wanted to develop exhibits and programs that would appeal to this audience.

Based on the success of the Latino Audience Development Initiative, the Vietnamese initiative used an outreach model that involved a three-phase process:

- community assessment and relationship building;
- development of an operational strategy, an exhibit, educational programs, an event, and marketing and governance strategies; and
- full-scale implementation of developed strategies.

From the outset, the initiative brought in advisers from the Vietnamese community to build long-term relationships and to help with exhibit and program planning. "We held focus groups to find out what was important to Vietnamese visitors," says Jenni Martin, director of education. "We learned about some cultural icons and discussed the pros and cons of having the labels translated into Vietnamese. Throughout our collaboration, the welcoming message that we sent was very important."

The Community's Perceptions

An analysis of the data from the focus groups shed some light on what the Vietnamese look for in their leisure destinations and how the Children's Discovery Museum did—and did not—meet their needs. Many Vietnamese parents saw a number of positive aspects to the CDM experience, including:

- A safe, clean environment
- Important focus on math and science
- Excellent customer service and friendly staff
- Valuable exhibits for younger children
- Genuine efforts to reach out to the Vietnamese community

However, focus group participants also pointed out many barriers to visiting the museum, such as the cost of admission, lack of transportation, parking fees, and the location. More specifically, many first-generation respondents were not comfortable with the location of the museum, which is not close to areas of high concentration of Vietnamese, making the neighborhood less familiar. They also found the logistics of paying for parking challenging. Furthermore, the lack of Vietnamese-speaking staff, particularly at the museum entrance, made it difficult for some families to communicate, contributing to their lack of comfort.

It also appears that perceptions of museums were a barrier. The word "museum" seemed to carry negative connotations for a lot of families. Respondents saw museums as passive, old, and academic versus interactive and engaging. In their minds, museums were associated with places that display old historical artifacts for visitors to view but not necessarily touch and interact with. Many focus group participants did not see how CDM provided more educational and fun experiences; in some cases, they weren't even sure what the goals of the museum were, despite having visited the museum before participating in the focus group discussions.

Values related to education more broadly may have played a role in these perceptions. Traditionally education is highly valued in Vietnamese culture and is perceived as being the sole responsibility of the school system and the teachers. Parents tend to keep some distance from their children's education. In addition, to some extent play and learning are seen as two distinct activities. This perception may be one of the reasons that focus group participants were not clear on the goals of the children's museum, which is intended to be both fun and educational.

Generational differences in the Vietnamese community also emerged. First-generation members, or those born outside the United States, tend to speak Vietnamese in the home and tend to live in more insular communities. They value their cultural traditions and enjoy sharing and talking about their memories of life and traditions in Vietnam.

Individuals who immigrated to the United States and children (referred to as 1.5 generation) and second-generation members (those born in the United States) are more likely to be acculturated, may speak the Vietnamese language but have limited reading and writing abilities, and in general are less tied to Vietnamese customs. They enjoy seeing their traditions reflected in their community and like the idea of exposing their children to the traditions. However, they also valued multicultural perspectives and sought to instill in their children respect for all cultures.

Planning an Exhibit for the Vietnamese Community

The first major project for the partners in the initiative was to plan a museum exhibit on mathematics and science called Secrets of Circles. The goal of the exhibit was to introduce young children to the concept of a circle as a geometric shape seen in nature and their everyday life. The exhibit included stations at which visitors can use a compass to draw circles; explore the rolling and spinning patterns of three-dimensional circles; and observe spinning circles that change into cylinders, a sphere, and a torus. Throughout the exhibit, children and their caregivers learn about the math, science, and beauty of this shape.

Based on feedback from the community and their own research, museum designers incorporated some key Vietnamese cultural icons into the exhibit. For example, bamboo was selected as the main building material for the exhibit, and the Vietnamese round boat and a rice sieve were used as examples of circular objects. Museum staff also deliberated about whether to translate the labels into Vietnamese. Despite their awareness that younger Vietnamese people may not read the language, they decided to move forward with the translations. "It was a good decision," says Martin. "In particular, first-generation Vietnamese were glad to see the text and graphics in their language."

According to the summative evaluation of Secrets of Circles completed by Sue Allen and Associates, many of the exhibit's elements succeeded in helping families feel more comfortable at the museum. Many visitors especially enjoyed seeing the round boat, which sparked conversation about their lives in Vietnam: "The round boat reminds me of the area where I used to live in Vietnam. This kind of boat is popular in the middle of the country. In the mornings, I used to walk to the beach to see the fish, shrimps, or crabs unloaded from these boats. The bamboo, the pulley, and the rice sieve on the wall all remind me of the good times in Vietnam."

Other visitors, however, were concerned that too many Asian elements were incorporated into the exhibit along with the Vietnamese ones. One visitor said that "the Circles exhibits should make it clear whether the theme is countries in Asia, like China, India, Laos, Thailand, Cambodia, or just the Vietnamese culture, when you have Chinese characters on the hats and Chinese lanterns and umbrellas."

Martin observed that the use of Chinese cloth hats turned out to be particularly problematic for Vietnamese visitors. "We started out with traditional Vietnamese straw hats, but they did not hold up, which made them a potential safety hazard," explains Martin. "Making the decision to change to the cloth Chinese hats had ramifications that we did not expect."

To address some of these criticisms, the museum is already working to improve the exhibit. They have purchased a traditional cyclo (or pedicab) to add as another example of a circle. They also are considering adding a Vietnamese drum.

It is interesting to note that much of the negative response to the exhibit, especially the inclusion of non-Vietnamese elements, came from first-generation Vietnamese. Generation 1.5 and second-generation Vietnamese were much less particular about those issues and were very enthusiastic about the exhibit. One community leader felt that despite these problematic details, the exhibit captured the essence of what she considered to be Vietnamese: "I love the look of it, coming in to the bamboo makes it really comfortable. . . . Sometimes science exhibits are more professional or academic, and less inviting. But this one with the umbrellas, it's a really fun place to be in. And it reminds me of Vietnam, just the different bamboo that I've seen in my life, it makes me really comfortable. And the fabrics and colors feel very rich."

Progress Made, More Work Ahead

From the outset, the initiative brought in advisers from the Vietnamese community to build long-term relationships and to help with exhibit and program planning. The evaluation of the initiative indicated that museum staff have developed very strong and solid relationships with community advisers. Advisers noted that they felt the partnership was a positive one, in which everyone's ideas were heard and which gave them an opportunity to share their knowledge and experiences. What's more, the advisers expressed great appreciation for being invited to participate and partner with CDM.

The strong relationships forged with advisory members have resulted in a cadre of people deeply committed to the mission of CDM and the vision of better serving the Vietnamese community. These advisers mentioned that their ongoing involvement emerged from the museum staff's commitment to diversity, manifested in the open, collaborative way they worked with the advisers.

While relationships with the advisers are strong overall, the most active and supportive partners were those who worked at organizations whose mission closely aligned with that of CDM. These partners not only understood what CDM offers, but also noted that their own organizations are working toward similar goals, such as education; as a result, these organizations were invested in the project. Because of the crucial role that partners play in the initiative and the fact that many are already stretched in terms of time and money, advisers commented on the need to expand community relationships beyond the current team.

The experience working on the exhibit and the initiative as a whole has been an eye-opening one for the museum staff. For one thing, the staff discovered that developing an understanding of and competence with a culture is an ongoing process. In fact, according to the Garibay Group's final evaluation, even after working on the project for several years, many staff members still felt tentative about their decisions and were concerned that they may inadvertently offend Vietnamese community members. One recommendation made by the evaluator that may help considerably is to hire Vietnamese staff who can serve as "cultural translators" for the museum staff who are not Vietnamese, helping to bridge the gap between the museum's culture and that of the Vietnamese community.

Although staff members learned a lot from the initiative, they recognize they still have a long way to go. "Being involved in the Initiative has raised many questions for me," says Martin. "I'm still not completely satisfied that we have been successful in our work with the Vietnamese community. We would like to continue to build that relationship."
(Based on an interview with Jenni Martin and on the following reports: "Secrets of Circles: Summative Evaluation Report," prepared by Sue Allen, Ph.D., Allen & Associates, and Children's Discovery Museum of San Jose: Vietnamese Audience Development Initiative, Garibay Group, Fall/Winter 2008)
[END OF CASE STUDY 7-1]

This case study illustrates the value of drawing on participants' cultural practices when designing informal learning environments. This can be accomplished by incorporating everyday language, linguistic practices, and local cultural experiences. While designers of informal programs and spaces for science learning have long recognized the importance of building on participants' prior knowledge and experiences, the integral role of culture in shaping knowledge and experience is not always appreciated. There are many challenges to forming true collaborations resulting in programs, exhibits, and activities that integrate traditional knowledge, beliefs, and practices with the knowledge and practices of Western science. However, the CDM's Vietnamese initiative demonstrates that success is possible (see Box 7-1 for further discussion of building community partnerships).

Indeed, research and evaluation on other efforts in museums to better address diversity show that the resulting enhancements can improve learning. For example, bilingual interpretive labels in English and Spanish allowed adult members who were less proficient in English to read the labels and discuss the content with their children, directly increasing attention and improving learning outcomes.[6] In another case, providing a Spanish-speaking mediator promoted more scientific dialogue. Finally, in a science program that was offered in English and Spanish, participating girls were very positive because they learned science terminology and concepts in both languages and thus could better communicate with their parents (who were predominantly Spanish-speaking) about what they were doing and learning. This increased their confidence and helped bridge the program and home environments.[7]

INCLUDING PEOPLE WITH DISABILITIES IN INFORMAL SCIENCE EXPERIENCES

Another group that is often excluded in informal science settings is people with disabilities. With the number of people with cognitive, physical, and sensory disabilities currently making up a significant portion (18 percent) of the population, this group also needs to be considered in the planning and development of informal science experiences.

People with disabilities face multiple obstacles when trying to take advantage of these opportunities. Some issues are physical; for example, navigating a space can be problematic for people in wheelchairs and for those who are blind. Other issues, however, are related to a culture gap that must be bridged, much like cultural differences between various ethnic groups and informal science settings. Removing cultural barriers, however, is much more difficult than addressing physical ones.

The following case study explains how designers at the Museum of Science, Boston, went about this task as they planned and developed an exhibit called *Making Models*. As planners at the Children's Discovery Museum did, Museum of Science staff worked closely with members of the targeted communities to make the experience both accessible and equitable.

[CASE STUDY 7-2]

Culturally Relevant Exhibits for People with Disabilities

The Museum of Science, Boston, has a long-standing commitment to developing exhibits for people with disabilities. More than 20 years ago, Betty Davidson, a museum exhibit planner who was in a wheelchair herself, paved the way by redesigning a diorama exhibit with multisensory components. Christine Reich, manager of research and evaluation, drew inspiration from that early work as she designed *Making Models*. The goal of this exhibit is to explain what a model is, present examples of different models, and give visitors the opportunity to experience how to make models. Their hope was to ensure not only that people with disabilities would have access to the exhibit, but also that they would be able to learn the science behind making models, largely because the material was presented in a culturally sensitive way.

Reich and the other members of the *Making Models* team set the bar high. They wanted to create some exhibits for people with many disabilities: wheelchair users, those who are blind or have low vision, and people who are deaf or hard of hearing. To accomplish this goal, they organized a community advisory group that consisted of people with various disabilities who were also experts on access, representatives from state agencies, or activists in the field. One member of the group, a science illustrator, had some expertise about modeling and also had multiple sclerosis. Another member had low vision and worked at a community services organization for older adults with low vision. Another, who was in a wheelchair, could move only his hands; this individual had extensive knowledge about psychology and the arts. Each advisory group member brought a much-needed perspective to the conversation.

The elements in the exhibit ended up incorporating many of the ideas discussed by the advisory group. For example, the human models were not just of able-bodied people. One of the male models was a tall African American with a prosthetic leg. The leg shown was not state of the art, either; it was the kind of prosthesis that ordinary people would probably purchase. And three models of hands showed them signing the letters A, S, and L, which stand for American Sign Language.

Interactives also were a part of the exhibit, and the key to designing them was to ensure that visitors could access them using multiple senses. "At the model-making station," explains Reich, "people with limited reach could create a model using beaded metal chains on a magnetic board. At another station, they could build a model by pressing buttons."

Two particularly innovative options allowed visitors to build models using light or sound. On a stage, visitors could manipulate color, the position of light, and its intensity to create a seasonal image, such as a sunset in winter or a sunrise on a summer day. The buttons and knobs that manipulated the light were easily reachable without moving, and there were places where visitors could rest their wrists.

At the sound station, visitors could select sounds from a series of electronic files to create a scene. Sounds included snoring, meowing, an alarm going off, or people chatting. Like the light stage, the sound models were created by pressing buttons and turning knobs.

Throughout the exhibit, visitors had access to audio and text labels, so learning was possible through either mode. The availability of multiple modalities for learning also meant that a sighted visitor could explore the exhibit with a friend with low vision, or that parents could have different ways to explain ideas related to the science to their children. The exhibit area also was easy for individuals in wheelchairs to navigate.

The Impact of the Exhibit

Did these adaptations increase the ability of disabled visitors to engage with the exhibit and to learn the science? According to the summative evaluation report, in many ways, they did. For example, those with mobility impairments—wheelchair and scooter users and amputees—could get around without any trouble. One obstacle reported, however, was that objects in a case were hard to see, and an amputee noted the need to have more places to sit down.

Blind and low-vision visitors, however, did find some parts of the exhibit difficult to access. Some expressed disappointment that they couldn't touch the objects described in the audio, while others were frustrated if they had trouble getting the sound to work. One blind visitor suggested the following: "The exhibit needs an overall orientation, and a Braille map would be helpful, too. Some of the stations need to provide more feedback to blind visitors in order to be accessible. . . . Some type of clearer pathway would benefit some disabled visitors."

The report also revealed that even though it is extremely difficult to make every exhibit accessible to every visitor, enough options were available, making the experience equitable in the opportunities it provided for learning. According to the evaluation report, about one-third of these visitors said their understanding of models changed as a result of the exhibit, a response rate similar to that of able-bodied visitors. Yet there was still room for improvement.

"The goal is to make sure that there are enough experiences so that all visitors feel included," says Reich. "And some exhibits carry more weight than others. If people are excluded from 'landmark exhibits,' they feel like they missed out on the experience."

Moving forward, Reich notes that many museums, including the Science Museum of Minnesota and the North Carolina Museum of Life Sciences, are working hard on issues of accessibility and equity. But there is much to learn. "Professionals want a checklist, a list of items they can check off and then say that they have done everything right," says Reich. "But that's not the way this works. What is really involved is a willingness to engage in a process of involvement and engagement, a change in mindset, and a re-assessment of what is 'normal.' Then people will realize that they need to tend to all these issues in order to reach everyone." *(Adapted from the Summative Evaluation Report, Program Evaluation and Research Group, and an interview with Christine Reich)*
[END OF CASE STUDY 7-2]

INTEGRATING NATIVE AMERICAN CULTURE WITH SCIENCE

In our discussion of the importance of culture in science learning, we have focused on how informal learning institutions can partner with members of the community, particularly those who represent nondominant groups, to rethink the way the institutions approach designing programs, exhibits, and other activities.. When successful, these kinds of initiatives integrate elements drawn from the nondominant culture with scientific ideas and practices and offer access points to science that may previously have been unavailable to members of the group.

The role of culture and the need for collaboration are particularly important when the beliefs, language, and cultural practices of a particularly group have historically been devalued or even suppressed. The experience of many Native American tribes provides one such example. Native Americans have long been disenfranchised from their land and culture, and they have even been discouraged from speaking their languages and carrying out traditional ceremonies. As a result, the value of native knowledge and their beliefs about the natural world have often gone unrecognized; in fact, many people perceive a conflict between native understanding of the natural world and scientific understanding.

The need to make science education meaningful for Native Americans has long been recognized by respected leaders in the field. Thirty years ago, the American Association for the Advancement of Science (AAAS) called for using an ethnoscientific as well as bilingual approach to teaching science in particular contexts. In response, scholars called for science education that directly relates to the lives of native students and tribal communities. Most scholars, including Glen Aikenhead, agree that to be most effective, learning environments must be connected and relevant to each particular Native American tribe.[8]

Native Science Field Centers, supported by the efforts of the Hopa Mountain program, strive to create such environments in their year-round programs for elementary and middle school students. These programs connect traditional culture and language with Western science. Currently there are three Native Science Field Centers—one on the Blackfeet Reservation in collaboration with Blackfeet Community College (Montana), one on the Wind River Reservation in collaboration with Fremont County School District No. 21 (Wyoming), and one on the Pine

Ridge Reservation in collaboration with Oglala Lakota College (South Dakota). The following case study focuses on the Blackfeet Native Science Field Center.

[CASE STUDY 7-3]

Merging Native Culture and Language with Science

During one of their after-school field trips, youth participating in the Blackfeet Native Science Field Center (NSFC) went out to gather willow. Before they began, the group huddled in a circle, recited a prayer in their language, and held hands while making an offering of tobacco. Helen Augare, director of the center, explained that the youth are learning that this is the respectful way to proceed before picking plants. By practicing this tradition, students learn that they have a reciprocal relationship with mother Earth and that they should take only what they need.

The participants then started their hike through knee-deep snow and thick brush to find and gather willow for their projects. Then they planned to travel back to their meeting place on the campus of Blackfeet Community College and use the willow to learn the process of constructing backrests and snowshoes—technologies that their ancestors engineered generations before them.

Activities such as this one are part of the Native Science Field Centers, whose overarching goal is to merge Western science concepts with traditional ecological knowledge of tribal communities. The program, launched in 2006, is held year-round, with four six-week sessions that run in concert with the seasons. During the school year, participants meet three times a week, and during the summer they come every day. The Blackfeet site is designed to provide technology, engineering, and mathematics learning opportunities for youth and adults by introducing them to culturally significant sites, birds, plants, and animals. Activities incorporate the language and offer learning enrichment through presentations by tribal elders and professionals.

The program is a community-wide effort. An advisory board ensures that program developers are implementing traditional knowledge in an appropriate way and provides guidance and support in developing cultural curriculum materials and finding resources. Parents, teachers, and tribal elders have contributed by donating materials for youth projects, sharing their knowledge, and volunteering their time during activities.

Buy-in for recruiting and retention is achieved during the program's orientation session for parents, who are amazed at how much is being accomplished. "We're trying to do more than just teach biology and ecology, and even more than just teach culture or history," Augare explains. "We're trying to show kids the spiritual element—how to take that in and make it a part of their world view."

A big part of the program is introducing participants to the land by monitoring sites and collecting data about culturally significant plant and animal species. "We went to tribal leaders to ask them what animals to include," says Augare. "Then we explain how they are part of the ecosystem, which they have a responsibility to care for."

To reinforce the importance of care for the land and the plants and animals that depend on it, the group worked with community members to put on a skit about climate change. A teacher fluent in the native language wrote the skit and helped the kids learn their lines—all in the Blackfeet language. The show emphasized how lessons can be learned from animal behavior and by observing the balance of the four elements—wind, fire, water, and land. Learning these lessons will allow the Blackfeet to adapt to climate change and keep mother Earth healthy.

Over the long term, the program is working to build an interest among native people in pursuing careers in science. With professionals from the community serving as role models, this generation has opportunities not available to their grandparents. Because of improved education systems and positive learning environments, there are a growing number of Native Americans studying science and selecting careers in different disciplines. More and more, native students feel proud of their heritage and celebrate the contributions to science made by their ancestors. They also are motivated to work toward the advancement of their tribal nation.

The Blackfeet program is still quite new, and its leaders are currently working on evaluation tools that reflect the indigenous perspective. Their goal is to be able to demonstrate how the spiritual connection can be a motivating factor in learning. "The Blackfeet are proud of their culture and proud of their history," says Bonnie Sachatello-Sawyer, executive director of the Hopa Mountain program. "This program, rooted in their values, will help give today's children the foundation they need to make informed decisions about their land and water when, as adults, they are called upon do so."

(Adapted from interviews with Bonnie Sachatello-Sawyer and Helen Augare)

[END OF CASE STUDY 7-3]

The success of the Blackfeet Native Science Field Center programs shows the potential power of informal learning experiences in science for engaging individuals from groups that are historically underrepresented in the field. In fact, several studies suggest that informal environments for science learning may be particularly effective for youth from historically nondominant groups—groups with limited sociopolitical status in society, who are often marginalized because of their cultural, language, and behavioral differences.

Evaluations of museum-based and after-school programs such as the Blackfeet Native Science Field Center suggest that these experiences can support academic gains for children and youth from nondominant groups. Programs and experiences that are successful often draw on local issues. Several case studies of community science programs targeting youth document participants' sustained, sophisticated engagement with science and sustained influence on school science course selection and career choices. In these programs, children and youth play an active role in shaping the subject and process of inquiry, which may include local health or environmental issues about which they subsequently educate the community.[9]

* * * *

Informal institutions concerned with science learning are making efforts to address inequity and encourage the participation of diverse communities. However, these efforts typically stop short of more fundamental and necessary changes to the organization of content and experiences to better serve diverse communities. Much more attention needs to be paid to the ways in which culture shapes knowledge, orientations, and perspectives. A deeper understanding is needed of the relations among cultural practices in families, practices preferred in informal settings for learning, and the cultural practices associated with science. The conceptions of what counts as science need to be examined and broadened in order to identify the strengths that those from nondominant groups bring to the field.

We highlight two promising insights into how to better support science learning among people from nondominant backgrounds. First, informal environments for learning should be developed and implemented with the interests and concerns of community and cultural groups in mind: project goals should be mutually determined by educators and the communities and cultural groups they serve. Second, the cultural variability of social structures should be reflected in educational design. For example, developing peer networks may be particularly important to foster sustained participation of nondominant groups. Designed spaces that serve families should include consideration of visits by extended families.

Things to Try

To apply the ideas presented in this chapter to informal settings, consider the following:

- *Think about how to design environments and materials that are compatible with different cultural groups you are serving.* For example, would it be helpful to design an exhibit or a program for one specific group, or would incorporating cultural icons into an existing exhibit be more effective in your setting? Would adding multilingual labels be useful for your multiple audiences? Would programs in other languages be important to offer?
- *Explore and nurture partnerships with local communities.* Determine which groups you want to work with and then invite representatives from these groups to partner with you to define goals and serve as advisers throughout the project. Cooperate or collaborate early to ensure true partnership on equal grounds. Allow yourself to question cultural assumptions.
- *Learn more about the cultural ramifications of learning.* Invite a local expert in this field to come to your venue to discuss how culture affects the work being done there. What do you need to learn about visitors to your setting? How can you make your environment more culturally relevant? Contract colleagues in your field who may already have garnered considerable expertise.
- *Be informed about and coordinate approaches with neighboring venues.* Contact nearby informal learning science environments to discuss their strategies for working with different members of the community. Can you work together to develop a joint program or activity that will be particularly meaningful to the different groups you are trying to serve?

For Further Reading

Aikenhead, G.S. (1998). Many students cross cultural borders to learn science: Implications for teaching. *Australian Science Teachers' Journal, 44*(4), 9-12.

Allen, S. (2007). *Secrets of Circles: Summative evaluation report.* Unpublished manuscript.

Calabrese Barton, A. (1998b). Reframing "science for all" through the politics of poverty. *Educational Policy, 12,* 525-541.

Garibay, C. (2008). *Children's Discovery Museum of San Jose: Vietnamese Audience Development Initiative.* Unpublished manuscript.

Karp, J., and LeBlong, J. (2004). *Making Models: Summative report, Museum of Science, Boston*. Unpublished manuscript.

National Research Council (2009). Diversity and equity. Chapter 7 in Committee on Learning Science in Informal Environments, *Learning science in informal environments: People, places, and pursuits.* P. Bell, B. Lewenstein, A.W. Shouse, and M.A. Feder (Eds.). Center for Education, Division of Behavioral Sciences and Social Science and Education. Washington, DC: The National Academies Press.

Web Resources

Children's Discovery Museum of San Jose: http://www.cdm.org/index.asp?f=0

Case Study: Making Models: http://www.exhibitfiles.org/making_models

Exploratorium: http://www.exploratorium.edu/

Hopa Mountain: http://www.hopamountain.org/nativeScience.html

Museum of Science: http://www.mos.org/

BOX 7-1
Key Steps to Building Relationships with Communities

If there is one lesson that can be learned from the experience of the Children's Discovery Museum, it is the importance of building strong relationships with communities of nondominant groups. The museum accomplished this goal by forming an advisory committee at the beginning of the project, and its assistance proved essential to the program. But even with the committee's guidance, subtle differences within the community, such as differences in attitudes between first and subsequent generations, were not recognized until after important decisions had already been made.

Other institutions have also begun their work with different cultural groups by starting at the community level. At the Exploratorium, for example, museum staff recognized how little they knew about both the Latino and Asian communities that visited the museum or could potentially visit. As a result, they set out to learn more these communities before doing any program planning.

In 2004, the Exploratorium began the learning process by going out into both communities to conduct informational interviews and recruit members for their advisory committee. Through collaboration with these leaders, the Exploratorium discovered that overcoming the language barrier is essential, along with developing a program that has some cultural significance. As the first step in reaching out to these two communities, the Exploratorium developed a series of public programs.

The first of the three, *Ancient Observatories: Chichen Itza,* used a compelling science topic as its starting point. The program was enriched through the addition of cultural activities and interpretation. It was conducted in two languages—English and Spanish. The next effort, *Science of Dragon Boats,* began with a cultural topic that was enhanced through the addition of science activities and demonstrations. The third program, *Magnitude X: Preparing for the Big One,* emphasized the relevance of a science topic to daily life and added activities and demonstrations to make this point. This program was conducted in three languages staggered over the course of the day. There was an English session, a Chinese one, and a Spanish one. "This was not easy to pull off," notes Garibay, who worked with the Exploratorium on their front-end evaluation. "It was an indication that museum staff took this work very seriously."

The experiences of both CDM and the Exploratorium point to several strategies that could be applied to other informal science environments. These strategies are summarized below.

- *Draw on cultural practices of the learners.* The language, practices, and experiences of visitors clearly affect their experience. By becoming aware of some of these practices, professionals in informal science can incorporate them into their settings. CDM had success with this strategy by incorporating cultural icons, such as the round boat, into its exhibit.
- *Develop bi- or multicultural labels.* Not only can labels translated in different languages provide specific content to diverse audiences, but they also can spark conversation and meaning-making, especially among intergenerational groups with varying language abilities. Garibay notes that bilingual labels allow adult visitors who were less proficient in English to read the labels and then discuss the content with their children, directing their attention to important features of the exhibit.

- *Build relationships with the community.* Working with community-based representatives from nondominant cultures is an essential part of the process. CDM's Jenni Martin notes the role that the Vietnamese community played throughout the initiative: "Working with the community is part of our mission as a children's museum," she says. "Leveraging trust with our partners, which include a community advisory group and the Vietnamese language media, has been critical to the success of our Initiative." Community leaders also can demystify museums for members of their community and help them understand the full range of available programs and activities.

8
Learning Through the Life Span

After visiting a science exhibit, a 73-year-old man, asked whether he learned anything, responded:

> You learn—it's amazing. …I'm going on 74 and …you're learning something new every day. And when you see a statement like scientists still don't agree about algae, whether they're plants. You know they work a little like a plant but then they don't and so some say, "yes it is" and some say "no it isn't." I'm looking at the spores—amazing tiny little specimens underneath the microscope—the variety. It's quite intriguing. I think anyone would find it interesting.

This man's keen interest in the exhibit and his perceptive response to it underscore one of the core values of learning in informal science settings—it is lifelong, occurring throughout the life span. What's more, learning takes place as people routinely circulate across a range of social settings and activities—after-school or community programs, clubs, museums, online venues, homes, and other settings in the community.

In developing exhibits and planning programs and events for learners, it is essential to know which segment of the population the experience is targeting and to plan accordingly. People's needs and interests change over the course of a lifetime, along with the way they process information and use tools, such as technology, to facilitate learning. They develop new interests and manage new tasks that arise depending on what stage of life they're in. Leaving school and entering the workforce, learning how to be self-reliant, becoming a parent and being responsible for children, and taking on the task of caring for aging parents are stages that often demand that people navigate and explore new scientific terrain.

Accordingly, motivations related to a particular aspect of science can shift over time. A young person who eagerly absorbs information about sea creatures simply out of a keen interest in the topic may, a few years later, be drawn to expand that knowledge in college in order to obtain the proper credentials for a career in marine biology. Decades later, this same person may be drawn into further study of ocean life simply for the pleasure of remaining current with up-to-date knowledge.

In this chapter, we explore some of the ways science learning varies with age. It is important to remember that, within these broad trends, individuals can differ tremendously. Their learning is influenced by prior experiences, gender, ethnicity, and other aspects of life that have nothing to do with age. And although the nature and extent of science-related learning may vary considerably from one life stage to another, most people develop relevant capabilities and intuitive knowledge from the days immediately after birth and expand on these in later stages of their life. In this sense, science learning in informal environments is truly a lifelong enterprise.[1]

CHILDREN AND YOUTH

At birth, children begin to build the basis for science learning. By the end of the first two years of life, individuals have acquired a remarkable amount of knowledge about the physical aspects of their world.[2] This "knowledge" is not formal science knowledge, but rather a developing intuitive grasp of regularity in the natural world. It is derived from the child's own experimentation with objects, rather than through planned learning by adults. In accidentally dropping something from a high chair or crib, for example, the child begins to recognize the effects of gravity. Although these early experiences do not always lead to accurate interpretations or understandings of the physical world, research has shown that these early naïve conceptions influence later science learning.[3]

As a child masters language and becomes more mobile, opportunities for science learning expand. Informal and unplanned discoveries of scientific phenomena (e.g., scrutinizing bugs in the backyard) are supplemented by more programmatic learning (e.g., bedtime reading by parents, family visits to museums or science centers, science-related activities in child care or preschool settings). Even in these initial years of life, children display preferences for some topics more than others. Such preferences can evolve into specific science interests (e.g., dinosaurs, insects, flight, mechanics) that can be nurtured when parents or others provide experiences or resources related to those interests.[4]

By the time children enter formal school environments, most have developed an impressive array of cognitive skills, along with an extensive body of knowledge related to the natural world.[5] It is also likely that they have become familiar with numerous ways of accessing scientific information other than through formal classroom instruction: reading, surfing the Internet, watching science-related programs on television, talking with peers or adults who have some expertise on a topic, or exploring the environment on their own.[6] These activities continue throughout the years in which young people and young adults are engaged in formal schooling, as well as later in life.[7]

It is also common for elementary schoolchildren to bring the classroom home, to regale parents with stories of what happened in school that day and involve them in homework assignments. These events help to alert parents to a child's specific intellectual interests and may inspire family activities that feature these interests. A child's comments about a science lesson at school may encourage parents to work with the child on the Internet or take him or her to a zoo or museum or concoct scientific experiments with household items in order to gather more information. In these ways, informal experiences can supplement and complement school-based science education.

To a considerable extent, children are dependent on others to provide opportunities for science learning—formal or informal. Especially in the early years of childhood, young people look to parents to provide access to reading materials, media resources, programmed environments (such as school, museums, zoos), and materials that can enhance informal science learning. Because of their limited knowledge base, children are also more dependent on adults to organize and interpret aspects of their learning experiences. However, children are also quite adept at creating learning opportunities from the resources available to them and deriving scientific insights from these opportunities, even at an early age.[8]

As young people move into adolescence, they tend to express a desire to pursue activities independently of adults.[9] This does not necessarily mean that relationships with parents grow more distant,[10] but young people do spend less time with parents or other adult relatives and more time with peers or alone.[11] Attachment to teachers also wanes across adolescence.[12] With these shifts, new opportunities for science learning become available that are not as closely tied to adult (especially parental) resources and activities. Young people gain greater access to school- and community-sponsored extracurricular activities (e.g., clubs or hobby groups) and, through part-time employment, may have more disposable income that can be devoted to hobbies or personal interests, including science-oriented activities.[13] They generally have greater mobility, especially with the advent of driving privileges that allows them broader reach into the surrounding community to pursue their personal interests. School systems tend to provide increasing levels of choice in course work with each advancing grade, allowing those with a penchant for science, mathematics, or technology to expand their exposure to science-related learning in formal contexts.

Community- and school-based programs are one setting that is receiving increasing attention as a support for science learning among children and youth. Programs, especially during out-of-school time, afford a special opportunity to expand science learning experiences for millions of children and youth. Out-of-school-time programs allow sustained experiences with science and reach a large audience, including a significant population of individuals from nondominant groups (see Box 8-1 for an overview of the history of out-of-school programs).

A range of evaluation studies shows that these programs can have positive effects on participants' attitudes toward science, as well as on grades, test scores, graduation rates, and specific science knowledge and skills. However, the body of research as a whole is difficult to make sense of because programs are focused on a variety of goals. Some place greatest emphasis on social or emotional well-being, such as developing positive attitudes, self-confidence, life skills, and social relationships. Others are more concerned with academic skills and improved academic achievement, as measured by standardized test scores, grades, graduation rates, and continued involvement in school science. Many are a blend of both. Because of these different emphases, along with the limitations of traditional academic assessments to measure the kind of learning that takes place in informal settings, it becomes difficult to draw definitive conclusions about the learning that is likely to occur. However, it is clear that the full potential of these programs for supporting science learning has not been tapped.

The following case study highlights one such program. Project Exploration is a nonprofit science education organization founded to address issues about access to science, particularly for populations historically underrepresented in the field. Project Exploration's programs run in out-of-school time settings, are free for participants, and specifically target girls and minority youth from the Chicago public schools.

[CASE STUDY 8-1]

Science at Yellowstone: An All Girls Expedition

Victoria, Kathryn, Xochitl, Latrise, Kassandra, Rana, Arieshae, Mystica, and Jasmine--participants in the All Girls Expedition to Yellowstone--began their experience with preparation in the classroom. Working with scientists, the girls learned about geology, water chemistry, and the wildlife at Yellowstone National Park. They also were introduced to the instruments that they would be using at the park.

For example, under the supervision of physical chemist Melanie Schroeder, the girls practiced using a thermometer to measure and record water temperature and litmus paper to measure its pH. Biologist Joshua Miller helped the participants build a giant food web by stringing together cards of plants and animals, showing the interconnections among different parts of an ecosystem.

At the end of the week, the group participated in an event called the Send-Off. The participants presented what they learned to family and friends and received a certificate of achievement and a backpack of supplies. Then, with scientists by their side, they were off to Yellowstone for a week of science and adventure.

An early morning start didn't faze this group. Brimming with energy and excitement, they couldn't wait to arrive at their destination. Most of the girls had never been away from Chicago, and many were not used to the physical exercise that was central to the experience. But throughout the week, they all showed that they could master what was expected of them— using an infrared thermometer to measure the temperature of hot springs, measuring the pH of water in Lemonade Creek, hiking long distances, and using telemetry equipment to track radio-collared coyotes in the wild.

Observation also was a key element of the experience. The girls observed the microorganisms living in Abyss Pool and used a spotting scope to locate animals on distant hills. And after watching steam pour out of Old Faithful, they discussed what geothermal energy is and whether it can be harnessed and used as an energy source for homes and businesses.

Learning on Many Levels

While the focus of the trip was on science, the girls learned much more than how to use scientific instruments and record their findings in their journals. They pushed themselves hard and learned that they could survive. They tapped into new skills as they developed strategies for working with their peers. As Kassandra, a ninth grader on the trip said, "What does [the trip] mean? It means girls working together and learning new things and helping one another along. It means proving yourself to those who think you're wrong."

For many girls, having an opportunity like this one was something they never thought possible. Mystica, an eighth grader, remarked, "Before this, I have never thought of going to Yellowstone or traveling this far away from home. But now I know that this is one of the best experiences of my life."

This "best experience" has proven to have had a lasting impact on many of the participants. Data gathered by Project Exploration show that about 43 percent of all girls who graduate from high school as a Project Exploration field expedition alumna go on to major in science in college. In addition, these girls become science majors at 5.3 times the national average rate. These findings reflect the goals of executive director Gabrielle Lyon, who points out: "We know that science today does not represent the diversity that is America. Project Exploration is working to change the face of science, literally, by creating and sustaining programs designed to not only *get* students involved with science but also *keep* students involved with science."

Perhaps the All Girls Expedition has had success in keeping girls involved because it looks at the importance of building self-esteem while also tapping into the girls' growing interest in science. Twelfth grader Latrise captures the essence of the science learning, as well as the camaraderie and mutual respect that emerges from this program: "We are smart, intelligent, young women from all over the city of Chicago yearning to explore the world of science and biology." Victoria, a ninth grader, echoed those sentiments, saying, "The two things I am most proud of are learning how to use a scope and how to track coyotes. I am very proud of these things because I mastered something new I have never done before."

"Letting people see how science unfolds is a terrific way to inspire students and get them excited about science," concludes Lyon. "This is just one part of our ongoing work to personalize science by focusing on the people who do the science and the questions they ask." *(Adapted from the Project Exploration Web site at http://www.projectexploration.org/summer-scrapbook-07.htm and input from Gabrielle Lyon, executive director of Project Exploration, and from Project Exploration program)*
[END OF CASE STUDY 8-1]

Project Exploration balances dual goals: promotion of positive social and emotional goals for the girls and supporting science learning. The opportunities to work with unfamiliar scientific instruments, observe animals in the wild, and engage in scientific experiments and to be successful in these endeavors not only resulted in science learning but also boosted their self-confidence and sense of their own competence. The program is also consistent with the emerging needs of adolescents for greater independence from their parents and more interaction with their peers.

By looking at the program in terms of the strands, it becomes clear how it provided a multifaceted science learning experience. The program succeeded in getting students involved in science by providing a hook—a trip to Yellowstone that included interesting hands-on experiences based in the real world (Strand 1). By learning how to use such tools as thermometers and tracking scopes (Strand 5), students increased their understanding of scientific concepts (Strand 2). The girls also were expected to write down their observations and the results of their experiments; both of these activities help in the development of scientific reasoning skills (Strand 3). By discussing their experiences and writing down their thoughts in the form of poems, narratives, and drawings, the participants also revealed that they were reflecting on the expedition and highlighting what they learned (Strand 4).

What's particularly compelling about this example is evidence that the girls themselves recognized their own accomplishments. They are proud of themselves, especially since many never thought they would have an opportunity to travel and have these experiences, let alone succeed at them. The girls' ownership of their success lays a foundation for future endeavors in which they are willing to take risks and try new things in college and into adulthood.

INFORMAL SCIENCE LEARNING EXPERIENCES FOR ADULTS

As individuals move into adult roles, they usually reserve a reasonable amount of time for leisure pursuits. Those with hobbies related to science, technology, engineering, or mathematics are especially likely to continue with intentional, self-directed learning activities in that area.[14] Science learning may also continue in more unintentional ways, such as watching television shows or movies with scientific content or falling into conversation with friends or associates about science-related issues. Some adults may focus especially on scientific issues related to their occupation or career, and in many cases their pursuit of scientific topics will be influenced by personal interests or (in later years) the school-related needs of their children.

Beginning in middle age and continuing through later adulthood, individuals are often motivated by events in their own lives or the lives of significant others to obtain health-related information.[15] Health-related concerns draw many adults into a new domain of science learning. At the same time, with retirement, older adults have more time to devote to personal interests. Their science learning addresses long-standing scientific interests as well as new areas of interest.[16]

Adults differ from children in their interest in science and in the way they approach different learning opportunities. Most adults become interested in a science topic because it has immediate relevance to their lives. Adults tend not to be generalists in their pursuit of science learning; instead, they tend to become experts in specific domains of interest in relation to everyday problems. The most obvious example is in the area of health. If an adult or a person close to him or her is diagnosed with an illness, such as cancer, that individual often goes to the library to take out books on the subject or goes online to find out as much as possible.

In some cases, this research could even lead to involvement in a support group. For example, one program for people with multiple sclerosis is a social club that also offers information about the administration of medication and the management of side effects. Patients whose treatments require regular injections are given anatomy lessons. To bolster learning and ease anxiety that may be associated with feelings of isolation, groups of patients may convene to share stories about their illness and treatment.

Sometimes, too, the everyday activities of adults lead to involvement in science. Adults may serve as chaperones for a school field trip to a museum, where they are put into the role of facilitator and are expected to answer questions, lead group discussions, and point out important aspects of exhibits. Over the course of a typical day, adults may notice a new kind of bird in the yard and take a moment to look it up online. Or they may tune into Science Friday on National Public Radio while picking up their children at school. If a particular topic piques their interest, then adults may seek out additional information online or even look into programs organized by informal institutions, such as museums, universities, science centers, and labs.

The move into adulthood can be both liberating and constraining in terms of informal science learning. On one hand, young adults generally exercise considerable control over their choice of activities and lifestyles. On the other hand, choices that they make may place constraints on their ability to freely pursue their interests in scientific phenomena. Certain careers or occupations will emphasize the need to master some scientific domains more than others. Parental responsibilities can trim the amount of free time available to pursue scientific interests, and parents may feel obligated to devote some of their leisure time to activities for their children. Therefore, science learning may be driven as much by the needs and interests of their children as by their own preferences.

Characteristics of Adult Experiences

Adults frequently perceive informal institutions and programs as geared toward children. In fact, there are many opportunities available for adults. Bonnie Sachatello-Sawyer and her colleagues surveyed more than 100 institutions that offer science learning experiences nationwide to assess the number and type of adult programs available. These researchers interviewed staff and participants from informal institutions of different sizes and types (art institutes, natural and cultural history museums, science centers, botanic gardens) offering different kinds of programs (credited versus noncredited classes, guided tours, lectures). Both studies found that 94 percent of all institutions offer some sort of adult programming, but most of the programs—63 percent—were designed for families or children.

This study also found that although most institutions are offering more adult programs than ever before, they are having trouble attracting and connecting with an audience. Part of the problem is with the kinds of programs offered. Lectures were offered more often than anything else, and they were viewed as dull from the adult learner's perspective. The adult learners told researchers that they were interested in programs that gave them exposure to unique people, places, and objects. They had positive impressions of programs that gave them access to new perspectives, attitudes, and insights.

In addition, the study found that no single teaching or facilitation method was more effective than another. The quality that participants were looking for in an instructor or facilitator was the ability to connect with the needs and interests of the learners and to help them discuss, integrate, reflect on, and apply new insights. In fact, for many participants, building a meaningful relationship with these facilitators was the most important part of the program. It is through such relationships that adult learners grow in their knowledge and understanding of a given topic.

As a way to explain the varying reactions adults have to programs in informal settings, Sachatello-Sawyer uses the image of a pyramid. Acquiring new knowledge and skill acquisition is at the bottom of the pyramid, followed by expanded relationships, which include more contacts in the community and new friends. At the next level, adults report increased appreciation of a topic as indicated by seeking out additional experiences and discussing the subject with knowledgeable individuals. Changed attitudes or emotions follow from the previous step, revealed through heightened self-confidence and taking the initiative to pursue new activities.

As adult learners reach the apex of the pyramid, they experience what Sachatello-Sawyer refers to as transformative experiences. Such experiences have caused learners to reevaluate their lives and make life-changing decisions, such as to leave one career for another or to find meaning in new experiences. For example, one learner reported that floating in the Grand Canyon made her realize that she "had a place in the cosmos and was part of the timeless nature of the canyon." While it is difficult to create such life-changing experiences, it is a goal to which program developers can aspire.

To illustrate an informal science program for adults and the learning that takes place, consider the next case study. Like the Cornell Lab of Ornithology's birdwatching program (Chapter 2), the case study also is an example of citizen science. But instead of focusing on gaining insight into animal biology and behavior, the purpose of this program is more practical: it is to track wildlife that crosses Highway 3, a busy road that cuts through the Rocky Mountains in Alberta, Canada. With greater awareness of the animals' habits, the possibility of developing interventions to reduce the number of wildlife-vehicle collisions increases.

[CASE STUDY 8-3]

Science in the Mountains: Road Watch in the Pass

At the Crowsnest Pass on Highway 3, it is not unusual to see bears, elk, cougars, and sheep ambling across the road. Before the highway was built, these animals claimed this route as part of their habitat. Now the animals must share the road with humans. The challenge for people is figuring out how this can be accomplished safely.

That's where the program Road Watch in the Pass comes in. Developed under the auspices of the Miistakis Institute for the Rockies, the program invites local citizens to share their knowledge about animal behavior. By providing information to the local community, their contributions can help prevent the 200 collisions that occur each year.

Participants have a couple of different ways to share their observations. The first is by making use of an online interactive map, which allows participants to plot the exact location of an animal sighting. This information is then sent to an online database, where it is stored. Although this approach has resulted in numerous observations (about 4,500 currently in the database), it also has some pitfalls. The observations are random, and there is no way to ensure that the entire road is being evaluated. Understanding accurately where wildlife cross the road and during which season is important for the development of strategies to reduce collisions along the road.

To address these concerns, the project added another data collection tool--the Road Driving Survey. It is designed for commuters who travel the same stretch of Highway 3 each day. Each commuter is assigned to the specific tract of the highway along which he or she drives on a regular basis. The commuters are given an electronic device, which enables them to key in what, if any, animals they see in real time. The advantage of this experimental design is that it becomes possible to evaluate more accurately where animals are crossing the highway and if there are seasonal variations in their movements. The data collected with the map and the survey complement each other, providing researchers with a more complete picture of animal behavior.

How can data like these be put to use? After four years of data collection, Road Watch information was used to develop a community map displaying wildlife-vehicle collisions, highlighting where collisions with wildlife are common. Based on June 2007 observations of bighorn sheep, the Fish and Wildlife Division of Alberta Sustainable Resource Development is using conditioning techniques to encourage them to stay off the road. The goal is to reduce the number of mortalities, especially near Crowsnest Lake. Rob Schaufele, coordinator for the project, notes that the data being collected also are being used by several other agencies for planning purposes. "The community of Crowsnest Pass should be proud of their contributions to Road Watch," says Schaufele. "It increases public and decision-maker awareness."
(Adapted from information found on the Road Watch on the Pass Web site at http://www.rockies.ca/roadwatch)
[END OF CASE STUDY 8-3]

Citizen Efforts Lead to Learning

This case study illustrates the pragmatic nature of adult involvement in science. Involvement in this program was precipitated by concern about the dangers to people and wildlife as a result of traffic on Highway 3. It follows that the people with the most at stake would be more apt to participate. In addition, the impact of the program on animal safety is contingent, at least in part, on participant investment in the effort.

There is evidence that learning has occurred. The number of observations in the database—4,500—indicates that people are interested in the program and motivated to participate (Strand 1). According to the results of a 2007 Road Watch online survey, 85 percent of the respondents (43 individuals) indicated that their knowledge of wildlife-vehicle collision and movement zone patterns had increased. In addition, through participation in the program, people gained much experience recording their observations (Strand 3) and using scientific tools (Strand 5).

On the basis of preliminary analysis of evaluation data, it is difficult to say whether a community of learners has formed. People do, however, discuss the program with their friends, although attendance at scheduled Road Watch meetings and events is erratic, ranging from low to high without a consistent pattern.

The online survey also asked whether respondents learned anything else besides information about wildlife. Participants indicated that they understood the potential of their data to affect future land-use decisions to improve safety conditions for both wildlife and people.

EXPERIENCES FOR OLDER ADULTS

Older adults are a unique population with whom informal institutions are working more frequently. Their abilities, needs, and interests require special attention in order to create programs that serve them. Long-standing misconceptions about aging, especially about the likelihood of such problems as memory loss and cognitive decline, have affected the way programming for this audience has proceeded. To improve this process, researcher Ann Benbow has compiled a list of strategies that should be considered when planning programs for older adults. These include explaining the activity clearly, setting aside enough time to complete it, and relating the activity to real-life situations.

Later adulthood often liberates individuals from the competing demands of work and family roles, but it may impose other restrictions on learning activities if such activities are not well designed to accommodate maturational changes of this age period. Many older adults are dependent on public transportation systems to access such community resources as public libraries, museums, or community organizations and scholarly institutions in which learning opportunities reside.[17] Learning environments also must make accommodations for adults' physical limitations. Museum exhibits that require too much walking or too much reading, especially of fine print material, can limit older adults' participation.[18]

One of the advantages of being older is that people have cultivated an extensive experience and knowledge base. They have a long history of family life, work experiences, and leisure pursuits that can serve as a starting point for new learning. In addition, research completed by Guy McKhann and Marilyn Albert has revealed that, throughout their lives, humans continue to generate new neurons in the hippocampus regions of the brain and that new neuronal connections are constantly being formed in response to new life experiences. Their research presents biological evidence that learning is truly lifelong.

Another relevant finding from research is that knowledge of general facts and information about the world do not diminish with age; in fact, experience and life skills lead to a more comprehensive understanding of the world. Self-worth, autonomy, and control over emotions increase or remain stable with age. There is evidence to suggest that older adults regulate negative emotions better than young adults while experiencing positive emotions with similar intensity and frequency. Overall, it appears that older adults can achieve an improved sense of well-being by pursuing experiences that are meaningful and tied to emotional information.

On the negative side, researchers Fergus Craik and Timothy Salthouse have found that older adults do face a steady loss in fluid intelligence, or processing capacity. This decline can adversely affect the performance of everyday tasks and learning as a result of a weakened capacity for attention and various types of memory performance. Because older adults often face declines in hearing, vision, and motor control, deficits in fluid intelligence can appear exaggerated. Studies also have shown that the extra effort expended by a hearing-impaired listener in order to successfully perform a task comes at the cost of processing resources that would otherwise be directed toward remembering the steps of the task.

Decline in fluid intelligence could have an impact on older adults' ability to use the computer. Older adults make more errors and perform at a lower level than young people do. In addition, they demonstrate an inability to edit out unnecessary information. Because the baby boom generation will presumably continue to use the computer into old age, it is important that website designers keep these deficits in mind and make adjustments accordingly.

Nonetheless, the evidence indicates that older adults can benefit from informal science programs, especially if some of these issues are considered in the program design. One such program, called Project SEE, or Senior Environmental Experiences, is a partnership between Ramapo College of New Jersey, the Meadowlands Environmental Center, and regional aging community services, including the Bergen County Division of Senior Services. Its purpose is to increase interest in the environment among seniors by making it relevant to their lives. The next case study is a glimpse of this program.

[CASE STUDY 8-4]

Science for Seniors: Project SEE

Senior citizens from 32 centers throughout the state of New Jersey can learn about ecology from the experts—scientists from the Meadowlands Environmental Center who are involved in cutting-edge research of the Meadowlands ecosystem. But the scientists aren't spending time on the road visiting these centers. Instead, the two groups are connecting through videoconferencing technology.

Program designers are preparing four modules to use with older adults. The first two have been completed and have already been used in four centers—the United Senior Center Hackensack, the Secaucus Senior Center, the Clara Maass Continuing Care Center, and the Lyndhurst Public Library. Each module asks a timely question and then uses a variety of strategies to present information.

The goal of the first module, *Should I tell my children and grandchildren to eat the fish and crabs they catch?* is to educate this audience about the continuing dangers of eating contaminated fish. Using this information, the participants can further discuss how to address the region's ecological problems.

Each module includes three sessions. The first, which takes place at the center, introduces the topic through hands-on activities. The next two sessions feature a videoconference with scientists from the Meadowlands. For example, during the videoconference for Module 1, scientists point out species of particular concern. They also go over the fish and crabs listed in the state's advisory and the potential health effects of eating them.

The following day, the group participates in the third and final session. Also a videoconference, this session focuses on what seafood people can eat and safe ways to prepare it. The session ends with the seniors competing for prizes during a Marsh Jeopardy Game.

"I've enjoyed working with each and every senior I have met so far," says Angelo Cristini, project director. "They are interested, engaged, and fun loving. Although it may sound hokey, this experience has certainly shown me that you really are never too old to learn new things."

(Adapted from the Project SEE Web site at http://www.marshmemoirs.com/about.htm)
[END OF CASE STUDY 8-4]

Effective Strategies for Work with Senior Citizens

Programming for senior citizens is a new field of informal science. Although there is little empirical analysis of such programs, it appears that forming partnerships with local organizations and area networks of aging services is a good first step. In developing the program, incorporating knowledge about the adult learner into the program design will enable it to be more targeted to the needs of this audience. There also appears to be some benefit to using technology to enhance the learning experience.

The informal science community is increasingly interested in serving older adults more effectively. From the research that has been done to date, it is evident that special accommodations will have to be made. For example, the Meadowlands program used videoconferencing so that the seniors would not have to travel to the center. Depending on the nature of the program, other accommodations and adaptations may be needed.

COHORT EFFECTS

Thus far, we have focused on broad changes with age that have the potential to affect science learning. An underlying assumption in these descriptions is that children will "grow into" the characteristics manifested by adolescents, who likewise will eventually display the characteristics observed in adults. However, some of the differences that can be seen across age groups do not disappear as individuals age. Instead, they serve as markers of distinctions among generations. These are known as *cohort effects*, meaning that they are attitudes, traits, or behaviors that typify a group of people born during a specific period, and they tend to stay with that cohort consistently across the life course.

Cohort effects are related to the common life experiences of individuals born in certain time periods. The term has it roots in population biology and has relevance in epidemiological studies in which subsets of a population are studied in relation to their exposure to certain sets of risks that can affect medical conditions, such as heart disease and cancer.[19] Cohort effects are studied in sociology and economics in relation to organizational culture and value orientations in society. One classic study, for example, charted the attitudes and behavior of a group of young people in California who came of age during the Great Depression, tracing the impact of these dire historical circumstances—and the world war and period of prosperity that followed—on their behavior across adulthood.[20]

The delineation of cohorts is always somewhat arbitrary, although they may be marked by major historical events. In the United States, some common cohort groupings are Postwar/Depression, Baby Boomers, Generation X (born between 1965 and 1979), Generation Y (born between 1980 and 1999), and Millennials. Delineations may differ across cultures or societies.

One important way in which cohort differences are manifest among these groups is in their experiences with and attitudes toward technology. World War I and Postwar/Depression groups grew up without television, much more attuned to the oral medium of radio, which requires more personal visualization of people and events. Baby boomers had TV, satellites, and a man on the moon but no personal computers until they were well into adulthood. By the time the first cohort of Generation Xers became teenagers, the computer revolution had started. Late Generation X and all Generation Y children in the United States have always had access to a wide variety of technology. And Millennials, born between 1985 and the present, have come of age (and continue to do so) with a full range of the current technological tools—e-mail, the Internet, cell phones, text messages, and social networking. Such differences have great potential to affect science learning. For example, in the WolfQuest case study (Chapter 1), it was clear that children and teens had no trouble learning science in the context of a computer game; in fact, learning on this platform was very comfortable for them. It is questionable whether the same could be said of many baby boomers, especially the older members of this group.

It is not always clear how distinctive characteristics of an age cohort will be manifest in each of life's stages. Instead, informal science educators and program designers must be responsive to the general principle that the program needs of each age group will be determined by the interaction of the primary developmental features and demands of the group's life stage, as well as the enduring characteristics that mark the group's age cohort. In short, each generation of children, adolescents, young adults, and older adults will be somewhat different, modulating the general script of a life stage by virtue of the idiosyncrasies of their cohort.

* * * *

Across the life span, from infancy to late adulthood, individuals learn about the natural world and develop important skills for science learning. Over time, their needs and interests change, affecting what kinds of science activities they choose to pursue. The preferences of individuals are partially affected by the time period when they were born and the impact of world events on their overall life experience.

Because one of the core values of the informal science community is to provide science learning experiences throughout the life span, it is important for program designers to consider the audiences they are serving. In particular, programs for school-age children and youth (including after school) are a significant, widespread, and growing phenomenon in which an increasing emphasis is placed on science. Clearly, the needs of children in out-of-school-time programs are very different from those of adults. Keeping these needs in mind and planning accordingly will lead to richer learning experiences.

Things to Try

To apply the ideas presented in this chapter to informal settings, consider the following:

- *Develop an understanding of your audience before developing a program.* Ask such questions as the following in your front-end or needs assessment:
  - What is their background?
  - What are their strengths and weaknesses?
  - What is their interest, and what motivate them?
  - What learning goals are you trying to accomplish? What do your audiences already know and not know? What do they want to know? How do they want to explore and discover? What would be the best way to meet those goals?
- *Seek out partners from the community.* This point has been reinforced throughout the book. When planning activities for different ages, it may be necessary for one group to seek out another so that an effective program can be designed. Project SEE, for senior citizens, is an example of how a partnership among three entities—Ramapo College of New Jersey, the Meadowlands Environmental Center, and regional aging

community services—joined forces to offer this audience a unique science learning experience.

- *Create advisory councils or groups.* Many museums have established a variety of standing councils or groups that advise them on matters of particular relevance, ranging teacher to teens. Standing advisory groups (rather than ad hoc focus groups, for instance) may make it more likely that diverse voices are heard and appreciated.
- *Be aware of new research.* In programs for all three groups, new information is always emerging about how people learn. Try to become familiar with the research base and use new findings to inform program design and development. There is a considerable body of knowledge on adult learning and adult learners that is relevant to informal science education and learning
- *Consider your audience diversity.* Previous chapters considered cultural and linguistic diversity, differences in interest, motivation, knowledge and situated identity as factors to consider when providing informal science learning experiences. Age and physical ability are certainly important aspects that are part of the consideration of just how diverse informal audiences are. Culturally-oriented designs or universal design principles (that acknowledge differences in physical and mental abilities of visitors) are ways to help serve the multiple audiences of informal science setting.

For Further Reading

Benbow, A.E. (2002). *Communicating with older adults: A guide for health care and senior service professionals and staff.* Seattle: SPRY Foundation/Caresource Healthcare Communications.

Craik, F.I.M., and Salthouse, T.A. (Eds.). (2000). *The handbook of aging and cognition.* Mahwah, NJ: Lawrence Erlbaum Associates.

Lee, T., Duke D., and Quinn, M. (2006). *Road Watch in the Pass: Using citizen science to identify wildlife crossing locations along Highway 3 in the Crows Nest Pass of southwestern Alberta.* In C.L. Irwin, P. Garrett, and K.P. McDermott (Eds.), *Proceedings of the 2005 international conference on ecology and transportation* (p. 638). Raleigh: North Carolina State University, Center for Transportation and the Environment.

Lindberg, C.M., Carstensen, E.L., and Carstensen, L.L. (2007). *Lifelong learning and technology.* Paper prepared for the Committee on Learning Science in Informal Environments of the National Research Council. Available at: http://www7.nationalacademies.org/bose/Lindberg_et%20al_Commissioned_Paper.pdf.

McKhann, G., and Albert, M. (2002). *Keep your brain young.* New York: John Wiley & Sons, Inc.

Milwaukee Journal Sentinel. (2008, October 8). *Playing with science: World of Warcraft competitors are involved in more than child's play, researchers find.* Stanley I. Miller. Available at: http://www.jsonline.com/business/32438914.html.

National Center for Education Statistics. (2006). *Digest of education statistics: 2006 digest tables.* Available at: http://nces.ed.gov/programs/digest/2006menu_tables.asp

Project Exploration (2006). *Project Exploration youth programs evaluation.* Available at: *www.projectexploration.org.* [Retrieved July 31, 2007].

Sachatello-Sawyer, B., Fellenz, R.A., Burton, H., Gittings-Carlson, L., Lewis-Mahony, J., and Woolbaugh, W. (2002). *Adult museum programs: Designing meaningful experiences.* American Association for State and Local History Book Series. Blue Ridge Summit, PA: AltaMira Press.

Salthouse, T.A. (1996). Constraints on theories of cognitive aging. *Psychonomic Bulletin and Review, 3*, 287-299.

Schaller, D., et al. (2009). *Learning in the wild: What WolfQuest taught developers and game players.* In J. Trant and D. Bearman (Eds.), *Museums and the web 2009: Proceedings.* Toronto: Archives and Museum Informatics. Available at: http://www.archimuse.com/mw2009/papers/schaller/schaller.html. Accessed March 31, 2009.

Steinkuehler, C., and Duncan, S. (2008). Scientific habits of mind in virtual worlds. *Journal of Science. Education and Technology, 17*(6), 530-543.

Web Resources

Center for the Advancement of Informal Science Education (CAISE): http://caise.insci.org/

Project Exploration: http://www.projectexploration.org/

Project SEE: http://www.marshmemoirs.com/about.htm

Road Watch at the Pass: http://www.rockies.ca/roadwatch/about.php

WolfQuest: http://www.WolfQuest.org/;
http://www.archimuse.com/mw2009/papers/schaller/schaller.html.

BOX 8-1
The Evolution of Out-of-School-Time Programs

Out-of-school-time programs have a long history in this country. They first appeared in the 19th century, and, over the years, they have evolved and changed to meet different needs, purposes, and concerns. Mostly, however, they have served the important functions of providing a safe haven for academic enrichment, socialization, acculturation, problem remediation, and play. Out-of-school-time programs continue to serve these functions even as they have grown in size and scope. Some are focused on homework help and tutoring, and others are enriched learning experiences or time for nonacademic activities, such as sports or arts and crafts.

Over the past 20 years, out-of-school-time programs have experienced tremendous growth, largely attributed to increased federal support for such programs as the 21st Century Learning Centers as well as the entry of more women into the workforce, which has meant that a greater number of children need supervised care after school. Politicians, educators, and parents increasingly view these programs as a necessary component of public education. The increased funding for the 21st Century Learning Centers is an indication of their growing importance: their budget rose from $40 million in 1998 to $1 billion in 2002. In 2007, the House of Representatives voted to increase funding to $1.1 billion.

The number of children participating in such programs also has increased, with school-based or center-based programs being the most common. In 2005, 40 percent of all students in grades K-8 were in at least one weekly nonparental out-of-school arrangement. An advantage of these programs is that they have the potential to provide large-scale enrichment opportunities for all children, including those from nondominant groups and low-income schools. In fact, at the 21st Century Learning Centers, the typical profile of a program participant is an individual who is black, from a single-parent, low-income home, and on public assistance. Because the programs are reaching nondominant groups in need of services, they are well positioned to make a significant difference in their lives.

9
Extending and Connecting Opportunities to Learn Science

It's 7:00 pm on a Sunday evening, and you have just returned home from a long, full day at the local aquarium. Your family saw many exotic fish and read about their behaviors on signs posted near their tanks. You also watched an IMAX film that showed some of these fish in their natural habitats. On the way home, your daughter talked about the fish she has in her classroom at school, and your son described the investigations they have been doing for a science unit on oceans. Now that you are home and relaxing, your daughter wants to see more fish, so she asks to watch the Disney/Pixar film, *Finding Nemo*. Afterward, you decide to sit down and watch some television before going to bed. One channel is showing *The Life Aquatic with Steve Zissou*, a Hollywood film inspired by yet mocking the character of Jacques-Yves Cousteau, the great science filmmaker. Meanwhile, upstairs, the long-running news program *60 Minutes*, is on another channel showing a segment on vacationers diving into ocean waters to observe sharks up close and personal, as well as the consequences of invading their territories. This segment intrigues your son, so he goes to the *60 Minutes* website to see a long list of people posting their comments on the show's content in real time.

EXPANDING OPPORTUNITIES FOR INFORMAL SCIENCE LEARNING

As this example illustrates, science learning, especially informal science learning, is an ongoing and potentially cumulative process. The impact of informal learning is not only the result of what happens during a particular experience, but also the product of events happening before and after an experience. Interest in and knowledge of science is supported by experiences across many different informal settings, as well as in schools. Although it is important to understand the impact of informal environments, a more important question may be how science learning occurs across the range of formal and informal environments and how formal and informal educators can capitalize on these connections.

Informal science educators are recognizing the power of providing ways for participants to extend and deepen learning experiences. For example, working in the museum context, Schauble and Bartlett designed an extended trajectory for science learning by using the idea of a funnel to map the way exhibits were laid out in space.[1] The outer edge of the funnel served all learners and consisted of easily accessible, compelling, and loosely structured experiences. The second level of the funnel was a series of quieter, restricted areas called Discovery Labs. Learners who chose to continue to pursue the big idea in question could move into these spaces. For example, at the Dock Shop, participants could explore boat design, including the design of different types of hulls tested for carrying capacity and various sail types tested with a wind machine.

The deepest portion of the funnel was designed for repeat visitors, such as museum members and children from the local neighborhood. The activities in this portion of the gallery built on children's prior experiences in the museum, at home, and at school. Visitors could borrow kits that were housed in the museum and distributed through local libraries. These kits contained materials that allowed children to extend their explorations in more detailed, sustained studies and to send in their results to the museum through Science Postcards. Learners who wanted to pursue a particular topic in even greater depth might choose to come back for an extended visit or several visits or to seek out other related activities, such as reading books on the topic or watching relevant television shows.

Many institutions extend their learning opportunities through systems for lending visitors objects and interpretive materials, such as books, other printed materials, activity kits, or videos, for a period of time. Some, like Science North in Canada, have made sharing educational resources a two-way street: they allow visitors or customers to contribute to the pool of resources made available to others, by borrowing or buying such resources from visitors who may have developed them as they engaged in scientific pursuits or science education activities outside the institution. Many museums also are turning to other forms of media, particularly the Internet, as a means of extending a visit to the museum through online activities.

In fact, broadcast, print, and digital media can play an important role in facilitating science learning across settings. Educational programming, "serious games," entertainment media, and science journalism provide a rich and varied set of resources for learning science. Through such technologies as radio, television, print, the Internet, and personal digital devices, science information is increasingly available to people in their daily lives. Although television is still the most widely referenced source of scientific information for most people, it may be losing ground to the Internet. New media, such as podcasts, webinars, and blogs, can support learning by expanding the reach of science content to larger and more varied audiences. They can also be used in combination with designed spaces or particular educational programs to enhance learners' access to natural and scientific phenomena, scientific practices (e.g., data visualization, communication, systematic observation), and scientific norms (e.g., through media-based depictions of scientific practice). What's more, interactive media have the potential to customize portrayals of science by allowing learners to select developmentally appropriate material and culturally familiar portrayals (e.g., choosing the language of a narrative, the setting of a virtual investigation) on their own cell phone or other handheld device.

Many museums, too, are experimenting with ways to make use of cell phones as personalized interpretation devices. For example, the Liberty Science Center in Jersey City, New Jersey, with funding from the National Science Foundation, has developed a program called "Science Now, Science Everywhere," which allows visitors to dial a phone number to receive additional information about an exhibit. Visitors can go online and find the number in advance of their visit so that they are ready to call in as soon as they arrive at the science center. Information comes directly to each participating visitor's cell phone in the form of a voicemail message or as a text. See Box 9-1 for a discussion of efforts to extend museum offerings through media.

Currently, the science center is working on expanding the reach of cell phones. Soon visitors will be able to sign up for a weekly photo challenge. While at the center, they can take a photo of an exhibit highlight and post it online. The photos will be reviewed and judged, with a new winner selected each week. If a visitor would rather just take a photo and save it on his or her phone, that, too, is possible. Through links to feeds on their phones, visitors also will be able to receive headlines of science and technology news posted at the center. And over the next year or so, more exhibits will be accessible through cell phones.

"We're continuing to think of ways to use cell phones to enhance the interactive experience," explains Wayne LaBar, vice president, exhibitions and featured experiences. "Cell phones are proving to be a way to continue to engage people with exhibits at the center even after they walk out the door."

While there is incredible potential for enhancing science learning through opportunities to extend and connect experiences, it is important to realize that little is known about how people learn about a single content or domain area across different informal settings and different media formats. Designing studies that examine this cumulative development of knowledge of skill is difficult. To illustrate this, consider a child reading a book about dinosaurs at age 3. She may like the book and ask to read it many times. Sensing her excitement for dinosaurs, her parents may take her to a museum to see an exhibit on her fourth birthday. The parents may have also bought her several dinosaur models from a local toy store during that period. A television program on dinosaurs may air after the museum visit, providing more information. And, in the era of networked computing, the family may seek dinosaur information together on the Internet.

Tracking all of this activity and determining the individual and collective impact on the child's emerging interest, knowledge, and skill are quite challenging. In fact, while it seems important to understand the cumulative effect of various, loosely connected learning experiences and to identify the relative contribution of individual experiences, it may be even more important for science educators to understand and appreciate the interconnections and to take them into account when creating and delivering science learning experiences for their audiences. With an appreciation that people will experience many and varied opportunities to learn science over the course of a lifetime, educators can design individual experiences in a way that better supports the overall journey[2]. For example, a museum exhibit about dinosaurs may be designed to optimize learning during the visit, with learning gains measured immediately after the experience. A different approach would be to design the exhibit to better connect to previous experiences and generate questions for further exploration at home. The measure of success of such an exhibit would be the quality of the questions generated and the nature of the next step visitors take to pursue those questions once they leave the museum.[3]

LINKING FORMAL AND INFORMAL SETTINGS

There is a growing recognition that individual museum visits, dinner-table discussions, visits to nearby parks, online searches or TV shows have cumulative effect on learning that we don't yet fully understand. We do know, however, that informal experiences for learning science need to be recognized and leveraged as part of an individual's personal learning pathway in science. Fostering links between experiences in school and out of school is one important way to enhance science learning. These linkages can help children and youth understand that learning is not restricted to schools and that there are opportunities to engage with science all around them. See Box 9.2 for discussion of major programs of research exploring the links between formal and informal settings for science learning.

Although there is tremendous potential in linking formal schooling to informal experiences that occur outside of school, there also many barriers to overcome when forging these links. For one thing, the goals and objectives of informal environments like museums, zoos, parks, planetaria, etc. do not match perfectly those of schools. Schools still focus much of their efforts on imparting knowledge while informal settings place greater emphasis on interest, emotion, motivation and engagement. Another important difference between schools and informal settings is that schools face increasing pressure to meet accountability requirements that place a premium on students' scores on tests. These same pressures have affected informal settings to a lesser extent. As a result, schools and informal institutions may appear to hold very different goals for learning when in fact, both share a common interest in enriching the scientific knowledge, interest, and capacity of students and the broader public.

Clearly in order to more effectively support science learning across the life span, it is essential to consider how schools and informal settings can work together more effectively. Below we consider some of the major points of intersection between schools and informal settings, focusing on field trips, after-school programs, and professional development opportunities for teachers.

THE VALUE OF FIELD TRIPS

School field trips to informal environments have a long track record and there is an abundance of literature that helps teachers and informal science educators plan field trips.[4] A 1997 study by John Falk and Lynn Dierking showed that all elementary and middle school students, as well as adults, could remember at least one thing they learned on a field trip. Over the short term, however, there are mixed results about the impact of field trips on children's attitudes, interest, and knowledge, although the majority of studies do show some positive changes in the areas of knowledge and attitudes.

Much of the work that has been done is on the structure of field trips and how it can be improved to facilitate learning. The critical factors that have been studied are advance preparation, active participation by students in the program, teacher involvement, and reinforcement after the field trip. We describe each of these areas below.

Advance Preparation

The purpose of advance field trip preparation is to give students a framework for interpreting what they will experience during the field trip and pointing out what they should pay attention to during the visit. Pre- and post-survey studies and observations show that students concentrate and learn more from their visit if they have engaged in related activities in advance.

Surprisingly, advance preparation is most effective when it reduces the cognitive, psychological, and geographical novelty of the experience. With some preparation, researchers Carole Kubota and Roger Olstad point out, students spend more time interacting with exhibits and learning from their visits.[5] Many studies, however, have shown that although advance preparation is beneficial, teachers spend little time on it.

Active Participation in Museum Activities

A review of more than 200 evaluations of field trips by Sabra Price and George Hein indicates that the most effective experiences include both hands-on activities and time for more structured learning, such as viewing films, listening to presentations, or participating in discussions with facilitators and peers.[6] For example, children who had an opportunity to handle materials, become involved in science activities, and observe animals and objects were excited about the experience. Similarly, a review of earlier field trip studies—from 1939 to 1989—by John Koran and his colleagues showed that hands-on involvement with exhibits results in more changes in attitudes and interest than passive experiences.[7]

To help keep students engaged throughout their field trip experience, Australian researchers Janette Griffin and David Symington argued for the inclusion of structured activities in the field trip.[8] Observing 30 unstructured classroom visits to museums, they noted that very few students continued exploring the museum purposefully after the first half hour of hands-on activities. Instead, most students were observed talking in the museum café, sitting on gallery benches, copying each other's worksheets, or moving quickly from exhibit to exhibit.

While individual field trips differ dramatically in their goals and character, it appears that successful ones combine elements of structured or guided exploration and learning that are designed with the unique opportunities of the setting in mind. They also incorporate opportunities for students to follow their own individual agenda, for example by exploring on their own (or in small groups). While teachers and the host institution may have to show that the field trip connects to standards or is linked to school curricula, field trips are also a way to introduce students to lifelong learning resources in their community.

Teacher/Chaperone Involvement During the Field Trip

Although studies have consistently shown that classroom teacher involvement in field trips can be key to their success, during most field trips the institution's staff members and not teachers are usually responsible for making the connections between the exhibits and classroom content, and a variety of studies indicate that teachers tend to assume a passive and unengaged role during field trips. The evidence indicates that the more involved teachers are in both planning of the trip and the visit itself, the more likely that the activities will align with classroom curriculum and be viewed as valuable experiences by teachers. Not surprisingly, the more engaged the teachers are, the more students will learn. Since field trips are often akin to "outsourcing" expertise, and informal science educators are in fact expected to assume the role of instructor, teachers still need to remain visibly engaged in order for their students to sustain their own participation and engagement. Informal science educators also often need teachers to help with class management and crowd control.

Parent and teacher chaperones are an essential element of school field trips, often required to supervise students and often difficult to recruit in sufficient numbers. Depending on the nature of the field trip experience, chaperones (like classroom teachers) could assume a much-enlarged educational role, providing interpretation and instruction and focusing student attention where needed and when appropriate. Unfortunately, there is little evidence that chaperones are used in this fashion. In fact, when the California Science Center experimented with chaperone-led field trips, teachers did not make much use of the program and the initial research on the effectiveness of chaperones as field trip docents was inconclusive.[9]

While teachers and parent chaperones could be a productive resource for the field trip, there are many informal educators who caution and recommend that teachers and chaperones be used sparingly to avoid adult intervention in student learning. It is a fine line between focusing a students' attention and changing the experience from one of discovery to one of lecture and demonstration.

Reinforcement After the Field Trip

Although teachers intend to do follow-up after a field trip, they often end up just collecting and grading student worksheets that are given out during the field trip. Griffin's 1994 study of field trips taken by students in 13 Australian schools showed that about half of the teachers reported that they planned to do follow-up activities but only about a quarter actually ended up doing so.[10] In addition, few students expected to receive meaningful follow-up, perhaps indicating what they experience most frequently. Studies in Canada, Germany and the United States produced similar findings.[11]

One of the reasons that developing meaningful post visit activities is challenging is that the experience often does not align with the classroom learning program. As a result, follow-up activities could potentially disrupt the work being done in the classroom. Even when the field trip does align with work being covered at school, connections between the two experiences often are not made. What's more, when teachers do try to have a discussion after the field trip, often it involves little more than asking students if they enjoyed the experience. When well-designed examples of classroom follow-up have been documented, they are in fact associated with positive educational impacts.

TAKING FIELD TRIPS TO THE NEXT LEVEL

While most field trips may involve one structured activity and a half hour of unstructured time, the state of Maine has developed a unique field trip experience. Not only is the informal science program aligned with the school science curriculum, but it also gives students entrée to a state-of-the art facility, the Gulf of Maine Research Institute (GMRI), housed at the Cohen Center for Interactive Learning.

The following case study describes LabVenture!, the GMRI program that is available to all middle school teachers and their fifth and sixth grade students in Maine. To date, more than 10,800 students from 177 schools throughout the state's 16 counties have participated in the program. It is an example of an ongoing relationship between a scientific facility and the schools that allows students to work with scientific instruments and use the skills of science to answer a compelling real-world problem.

[CASE STUDY 9-1]

The Mystery of the X-Fish

When the Gulf of Maine Research Institute (GMRI) opened its doors in 2006, the dream of its founders was to offer as many Maine middle school students as possible the opportunity to experience real science. Through LabVenture!, their dream is slowly becoming a reality.

"We charter three buses and pick up kids as much as seven hours away to bring them to the Center," says Alan Lishness, LabVenture's director. "When they arrive, we take them into the theatre laboratory and show them an immersion film, where they zoom into their school, western Maine, the West Atlantic, and the Atlantic. It's insanely exciting. And believe it or not, in a room of 48 students, it's possible that as many as 80 percent have never seen the ocean."

The student's total immersion film experience is only the beginning of the day. The core of their adventure begins after the introduction, when they divide into teams of three or four to try to figure out what the X-Fish is. They do so by solving problems set up at four different stations.

Each station offers its own unique experience. At one station, students observe a dead fish, paying close attention to the size of its mouth and whether it has teeth. On the basis of their observations, they record a hypothesis about characteristics of the fish. At another station, students study the X-Fish's stomach contents to determine what it eats.

A third station shows students how to find the X-Fish, first on a scientific cruise in the Gulf of Maine and then on a fishing expedition. During the expedition, the team works together to make decisions, which determine how profitable the trip turns out to be. At the fourth station, the students come face to face with a large tank of fish. They observe the fishes' behavior and then imitate it by running around the tank. The trick is to never bump each other, just as schooling fish swim together without getting in each other's way.

After all the students have visited each station, they spend 20 minutes preparing a presentation. Each team presents their ideas about the X-Fish. Just as scientists do, the student scientists work together to solve the mystery.

The program doesn't end when the students and their teachers leave the center. They can continue to discuss the experience by accessing personalized student websites. The websites document the students' thoughts and ideas, which have been saved online throughout the day. Students can review and annotate their websites, as well as continue to interact with GMRI staff through the center's blog.

But perhaps the biggest bonus of the experience comes from observing the kids and how well they work together. "What comes across watching the kids is how they treat each other with respect and learn from one another," says Lishness. "We expect the world from them, and they rise to meet—and even surpass—our expectations."

(Adapted from an interview with Alan Lishness and the following report: Baldassari, C. (2008). LabVenture: At the Cohen Center for Interactive Learning. Summative Evaluation Report. Program Evaluation and Research Group, Lesley University)

[END OF CASE STUDY 9-1]

What Did the Students Learn?

Based on the summative evaluation of LabVenture!, much of the learning that took place was in the development of inquiry skills (Strand 3). Based on responses from an online survey, about 74 percent of the students in the research sample said they learned about conducting scientific investigations by observing, forming hypotheses, collecting evidence, and analyzing their results.

The second area of learning mentioned most frequently by the students was working as part of a research team (55 percent). In addition, about 50 percent of the students said that they had the opportunity to figure out how to use scientific tools. Both of these learning gains correspond to Strand 5.

The students also noted that their visit to GMRI piqued their interest in marine science. Nearly half (47 percent) wanted to understand more about the Gulf of Maine watershed, and more than one-third expressed new interest in local freshwater resources.

But equally important, the kids experience learning as enjoyable and satisfying. "I learned how much fun oceanography can be," one student says. Another mentioned learning about the different types of tools scientists use. And one student expressed his opinion succinctly: "I learned a lot of cool stuff that I didn't know."

From the teachers' vantage, the experience at GMRI reinforced the fifth and sixth grade science curriculum, which includes the study of weather, environmental sciences, ecology, and watersheds and estuaries. Also stressed in these grades is the development of scientific inquiry skills. In the view of many of the teachers surveyed, GMRI offers their students a chance to learn some of this content and practice science skills in an authentic setting. As one teacher put it,

"[GMRI] fits the curriculum like a glove....It goes perfectly with our Invertebrates unit on fish classification and is a great hands-on science experiment for my students."

Another teacher echoed those sentiments, adding that "my students would *never* be exposed to anything dealing with Marine Science otherwise (and I can say that for grades K-8); this program is a much-needed addition to our science curricula."

ANOTHER MODEL FOR LINKING SCHOOLS AND INFORMAL SETTINGS

The LabVenture! case study illustrates how a research institution can develop and sustain an ongoing relationship with local schools through what is primarily a field trip experience. Through this relationship, students have an opportunity to experience science in an authentic setting, using real scientific instruments.

The next example discusses a relationship between schools (local high schools), the city in which these schools are located, a large science center, and a nature center that goes beyond the field trip model. Instead, it involves a long-term, sustained experience that capitalizes on unique local resources. This collaboration has evolved into a positive learning experience for students young and old.

[CASE STUDY 9-2]

The Lake Washington Watershed Internship Program

In Washington State, a year-long program for high school students, called the Lake Washington Watershed Internship Program, is made possible through a unique collaboration among the city of Bellevue, Bellevue's five high schools, the Pacific Science Center, and the Mercer Slough Environment Education Center. Throughout the year, 27 students meet once a week to learn about the watershed, conduct hands-on experiments, and work to restore the creek beds around Mercer Slough.

One way the program recruits student interns is by going into the schools and seeing who is involved in after-school ecology clubs. After an interview process, the interns are selected. Many stay with the program for three years, from tenth through twelfth grade. During the first year, they participate as volunteers, but in subsequent years the students are paid.

Perhaps the most exciting part of the program is the opportunity that the high school students have to go into local elementary schools to teach younger children about the environment. "They create their own lesson plans and become really passionate about environmental education," says Julie Rose, the program coordinator. "I hear some of the kids say that the internship inspired them to go into teaching."

To learn more about the impact the program has had on the high school students, Rose and her colleagues posted a survey on the Internet through Facebook. Although the data are preliminary, it appears that interns stay in touch with each other and discuss how the program has affected their lives.

This past year, Rose and her colleagues reached the milestone of seeing more than 100 interns go through the program. As a testament to the program's value, the Pacific Science Center has just made it a permanent part of its budget. "The Science Center recognized that the program is worthwhile," explains Rose. "It is involved in the community and teaches people that science is fun and interesting."

[END OF CASE STUDY 9-2]

OUT-OF-SCHOOL-TIME PROGRAMS: AN OPPORTUNITY FOR PARTNERSHIPS

Another way that formal and informal science settings can join forces is to offer unique opportunities for students through out-of-school-time programs. Historically, relationships between schools and out-of-school programs—particularly community-based out-of-school programs—have often been characterized by mutual mistrust and conflict. In a report based on 10 years of research studying approximately 120 youth-based community organizations throughout the United States,[12] Milbrey McLaughlin explains:

> adults working with youth organizations frequently believe that school people do not respect or value their young people. Educators, for their part, generally see youth organizations as mere 'fun' and as having little to contribute to the business of schools. Moreover, educators often establish professional boundaries around learning and teaching, considering them the sole purview of teachers. If we want to better serve our youth, there is an obvious need for rethinking the relationship between schools and out-of-school programs, particularly for out-of-school programs that have an academic focus such as science.[13]

There are different models of relationships between schools and out-of-school programs.[14] At one extreme, there is the model of "unified" programs that are the equivalent of what is now called extended-day programming. Under this model, out-of-school programs can become essentially indistinguishable from school, since they take place in the same space and are usually under the same leadership (the school principal). At the other extreme lie "self-contained" programs, which intentionally choose to be separate from schools. Taking place in a different location, they often provide students with an entirely different experience from school.

Many programs operate between these two extremes. In some cases, the out-of-school curriculum is closely connected to the school curriculum. In such programs, the program coordinators and staff know on a week-by-week basis the material teachers are covering in class and can directly connect it to out-of-school activities. The result is that the out-of-school science experience is essentially an extension of school science, but with a more informal feel.

In other cases, the out-of-school science programs connect their activities to the general school science curriculum and standards but not to what students are learning in class on a daily or weekly basis. This approach avoids some of the conflicts between science in schools and out-of-school programs, while allowing out-of-school programs to support students' learning in schools. It also has logistical benefits, since it does not require the same level of planning and day-to-day communication between schoolteachers and out-of-school staff.

Finally, in some programs, out-of-school science is entirely disconnected from school science. Directors, coordinators, and staff of the programs make sure that participants are engaging in high-quality science experiences, but they do not consider it essential for students to connect out-of-school science to school science. In some cases, these programs may go so far as to argue that by keeping the two worlds separate, out-of-school programs can provide students with an alternate entry point into science if they have already been turned off from school science.

The Multicultural Education for Resource Issues Threatening Oceans (MERITO) program in Monterey, California, illustrates a middle ground, where the out-of-school time curriculum and activities are coordinated with classroom activities, but not necessarily in lockstep. The MERITO program is a collaboration among the Monterey Bay National Marine Sanctuary (MBNMS), local school districts, and other local stakeholders. Its purpose is to provide underrepresented students with hands-on field experiences and in-class activities to teach them about nature and to instill in them a desire to protect the habitat. The program has two goals: to reach the community's growing Latino population and to teach this population about the importance of protecting the area's pristine shorelines and marine life. It is funded in part by the National Oceanic and Atmospheric Administration's California Bay Watershed Education and Training Program. The next case study is a glimpse of this program.

[CASE STUDY 9-3]

Science by the Sea: The Monterey Bay National Marine Sanctuary and Pajaro Unified School District Working Together

In 2002, MERITO launched a pilot program in partnership with the Pajaro Unified School District. Working with Pajaro Middle School and the Elkhorn Slough National Estuarine Research Reserve, educators began by developing activities and field experiences. These experiences would become a full curriculum aligned with California state standards and designed specifically for a diverse population of learners.

During the first year, 34 lesson plans were piloted with middle school students. Lessons ranged from native plant restoration, to shark tagging, to crab monitoring. Students met once a week to work on these hands-on science activities.

But lessons were not the only element of the curriculum. As part of the program, scientists from the field visited the students in their after-school setting to share their research with them. To cap the experience, participants went on numerous field trips to such places as the Watsonville Waste Water Treatment Plant, the Monterey County Waste Management District, and the Monterey Bay Aquarium.

The pilot program was a success. As a result, the former school superintendent requested funds to expand the program to the three other middle schools in the district. Karen Grimmer, acting superintendent of the Monterey Bay Sanctuary and a champion of the program, summarized the reasons behind the program's effectiveness: "Our communities are multinational and multilingual in nature. Our programs need to reflect the community in order to successfully communicate the importance of protecting our coastal and ocean resources."

The program introduced students to the precious environment in their own backyard. Once students had a better understanding of this ecosystem and the role they could play in protecting it, they embraced the charge and became stewards and advocates of the environment. In the process, the students also learned important science concepts and became energized and excited about the possibilities available to them through a strong background in science. *(Adapted from the following sources: 2002-2003 Report, MERITO website at http://montereybay.noaa.gov/educate/merito/outreach-community.html and National Marine Sanctuaries News & Events: Innovative Education Program Heightens Ocean Awareness at http://sanctuaries.noaa.gov/news/features/0706_merito.html)*

[END OF CASE STUDY 9-3]

Documenting the Learning that Occurred

At the beginning of the program, students were given a pretest to see what they knew about the watershed. The results showed little knowledge of this environment. So the teachers began by introducing the students to the basics in these areas. Throughout the year, they built on that foundation in a methodical way. This approach turned out to have tremendous payoffs.

As the year progressed, the evaluation team observed that students were able to explain the connections between watersheds and oceans, how the health of local waters affects humans and wildlife, and why watersheds and oceans need protection (Strand 2). In addition, students were excited about what they were learning and brought their families to community events, such as for Earth Day (Strand 1). The students worked with their families to create and distribute posters on storm drain pollution.

By the end of the first year, the program could claim some success. Although they had limited resources, the partnership between formal and informal education played a pivotal role in introducing children to their environment and what they could do to protect it.

The Value of Collaboration

These three examples—LabVenture!, the Lake Washington Watershed Internship Program, and MERITO—illustrate the potential of collaborations between formal and informal settings to maximize learning opportunities for students. Educators involved with the informal science program became knowledgeable about the science curriculum so that they could provide the students with complementary experiences.

Because these programs take place outside school, they have the advantage of providing key instruction away from the pressure inherent in the formal school environment. These advantages could help reach students who have difficulty learning in school, are turned off by formal education, or are looking for a different kind of experience to inspire them to take their interest in learning to the next level.

PROFESSIONAL DEVELOPMENT IN INFORMAL SETTINGS

Informal settings have long been recognized as an ideal place for professional development, largely because of their emphasis on learner-directed learning in a phenomenon-rich setting. In fact, teacher professional development is offered extensively by informal institutions such as museums, science centers, zoos, education and outreach staff of parks, etc. for mainly three purposes: to provide content knowledge to pre-k to 12 teachers, to provide pedagogical skills based on informal instructional techniques, and to promote the use of teaching materials (often developed by the institution itself). Until recently, however, their role has been relatively undocumented, and much of the evidence for their effectiveness or even successful practice is hidden in evaluation studies that have not been made public. There is evidence that teachers make extensive use of professional development provided through informal institutions and that they enjoy the different perspective provided by informal settings. However, little is known about whether professional development provided by informal science settings is more effective than that offered by other providers,

David Anderson and his colleagues from the University of British Columbia, Canada, studied how informal science settings could be used for a pre-service program.[15] The setting selected was the Vancouver Aquarium Marine Science Centre. The program began with preservice teachers participating in a three-day intensive program, which served as an orientation to the aquarium's educational programs. They also learned about student-centered, hands-on pedagogy and the institution's educational goals, described as "developing inspiration, curiosity, and marine stewardship." Following the program, the teachers spent 10 weeks working in a school. Then they returned to the aquarium for another three weeks to work in the educational programs under the guidance of aquarium staff

After the school and aquarium segments were completed, Anderson conducted two focus groups with the teachers, analyzed reflective essays they wrote during the semester, and made ethnographic observations at the aquarium. Based on the teachers' reflections and experiences, the researchers determined the impact of the experience in terms of the their understanding of the big picture of education and their growing sense that learning can take place in many settings; their understanding of education theory, their classroom skills, sense of autonomy, commitment to collaborative work, self-efficacy, and recognition of the power of hands-on experiences in learning science. While based on self-reports of a relatively small sample, the results suggest that this is a promising way to integrate teacher education in formal settings with instruction in informal learning environments. Clearly, however, further research and development are needed to document these findings.

Existing research and a variety of evaluation studies suggest that teacher professional development offered by informal science institutions should adhere to the criteria below:

- goals need to be defined clearly and need to be obtainable;
- programs should be developed in collaboration with teachers and schools to ensure the applicability and usefulness of the strategies offered (conduct a needs assessment);
- programs ought to aim beyond the immediate professional development experience and focus on implementation in the classroom with attention to fidelity or implementation while allowing teachers to adjust to their specific situation;
- professional development experiences need to allow teachers to learn from one another, share experiences and model new strategies;
- online offerings need to include "practice at school"; and follow-up support should be provided.

Taking the Lead in a Statewide Initiative

In some instances, informal settings can take the lead in improving the quality of science education in formal settings. In the late 1990s, the Pacific Science Center in Seattle was instrumental in working with other stakeholders to implement a statewide systemic reform effort called LASER (Leadership and Assistance for Science Education Reform). Part of a strategic leadership team, the Pacific Science Center helped bring exemplary inquiry-centered science curriculum materials to the state's elementary schoolchildren. Along with the new curriculum materials, the leadership team also ensured that teachers received professional development before presenting the material in the classroom.

Many evaluation studies have been conducted on the LASER project. In 2004, RMC Corporation investigated the relationship between professional development and the number of fifth grade students meeting the standards on the state's science test.[16] The results showed a strong positive correlation. The evaluators also determined that students made significant gains in their understanding of science from pre- to post-assessment, which took place after the students had completed work on several modules.

The Pacific Science Center is unique in that it has the capacity to lead such a large-scale effort. It is well positioned to seek private funding, build a coalition of stakeholders, and galvanize community leaders and politicians to get involved. While many informal science institutions are not able to assume such a large role in a major reform effort, this example does indicate the invaluable contributions possible by well-established informal science environments.

LEARNING PROGRESSIONSAND PREPARATION FOR FUTURE LEARNING

Work on learning progressions[17] in science is an emerging area of research in science education that could be supported by the informal science community. A learning progression organizes the study of science so that learners can revisit important science concepts and practices over many years. For example, the big ideas of science, such as evolution and matter, are introduced during the early grades; as students' capabilities increases, greater depth and complexity about these big ideas are added. At each phase, learners would be able to draw on and develop relevant capabilities across the strands.

Informal science environments could play a complementary role in supporting the understanding of these key ideas. For example, a program or exhibit in an informal setting could be designed specifically based on our understanding of learning progressions. This could be done by ensuring that the activity or exhibit is tightly aligned with K-12 science curriculum goals. The New York Hall of Science, working with the Miami Museum of Science and Planetarium and the North Museum (a small natural history museum in Lancaster, PA) and collaborating with a developmental psychologist from the University of Michigan, is developing a traveling exhibition on evaluation that is based on current understanding of children's naïve reasoning. It is designed to lead children of various ages through a series of increasingly more complex understanding of pre-evolutionary concepts. While the results of the approach on children's learning are not yet known (the study is currently under way), the inclusion of a learning progression researcher fundamentally altered the design process and the goals for the exhibition: the museum experts were far more inclined to see smaller steps in individual understanding as success and were developing an exhibition that was geared at specific levels of learning progressions within a pathway to understanding key aspects of evolution.

Alternatively, informal environments could differentiate themselves from the K-12 agenda by going "broad" on issues that the formal community chose to go "deep" with. In this way, informal environments could bridge the gap in teaching and learning by providing information not included in the learning progressions.

Another promising new area of study is the concept of "preparation for future learning" which recognizes that learning experiences might not always immediately and directly lead to increased knowledge or understanding. Instead, they may prepare the learner by creating cognitive dissonance or other forms of mental preparation that enhance the learning success when the learner encounters a later opportunity to build on the original experience (like an explanation given by a parent, or a follow-up to a field trip in the classroom). This has implications for informal settings like museums, since the purpose of the museum visit on a school field trip may not lie in conveying specific knowledge, but to prepare, through original experiences, the learner for subsequent classroom instruction. In that sense, preparation for future learning serves as a reminder that informal and formal learning are all interconnected aspects of the same overarching principle: a quest for lifelong learning that allows everyone to explore the natural and build environment and grow in their knowledge, understanding, and appreciation of the world.

* * * *

Science learning has the potential to cut across many platforms. Interested learners can go to an aquarium to observe sea life, go home and find more information on the topic on the Internet, and watch a television program in the evening. As technology becomes more sophisticated, many ways to link museums and other designed settings to home computers are becoming available. People can already view some museum collections online, and podcasts and webinars make events held at different settings accessible to a wide range of learners.

The relationship between formal and informal environments is of particular interest; in fact, preliminary research indicates that each setting has much to offer the other, but determining strategies that are applicable to multiple environments is still under way. Based on the research, however, informal science institutions have a role to play as destinations for field trips, settings for out-of-school-time programs, and places where professional development activities are held.

Things to Try

To apply the ideas presented in this chapter to informal settings, consider the following questions:

- Developing a collaboration between formal and informal settings involves considering many issues. If you are interested in embarking on such a collaboration, consider asking the following questions:
    - Is there a shared vision? Do all stakeholders know what they want to get out of the collaboration? Have reasonable goals been established to help all involved realize their vision?
    - Is the informal setting committed to working closely with the schools to develop a program that works for everyone?
    - Conversely, are the schools committed to working closely with informal settings? Does each of the partners know about other partners' assets and constraints?
    - Have clear and consistent lines of communication been established? Have informal settings considered the best ways to talk with schools? For example, is e-mail better than phone calls? Are more frequent, brief exchanges better than less frequent, more involved encounters? Are there mechanisms in place to inform parents about the nature of the relationship and progress being made? Has there been a staff person assigned to monitor the relationship and be accountable for successes and failures?
    - Are teachers being sufficiently supported by the informal setting? Are strategies in place to build trust and establish a strong relationship in which teachers and museum staff are learning from each other?
- This chapter has explored ways to strengthen the connections between formal and informal environments, but it is clear that more research is needed. If possible, consider how your institution could contribute to the research base. Can you set up

studies that explore how people routinely traverse settings and engage in learning activities across the board, from formal settings to informal ones?

- Technology may open the doors to greater access to science learning to wider, more diverse audiences. Has your institution developed ways to use technology to expand its reach? Using the ideas in this chapter, consider how technology can be used in your setting to help extend science learning, but also how to use technology to integrate school and out-of-school learning experiences.
- When developing programs and materials that connect formal and informal settings, ensure that the needs of each side are known and that programs or materials are developed with sufficient early inputs by each stakeholder. Packaged field trip experiences or curricular materials should be developed in close collaboration with teachers and students and pilot-tested before implementation, and the benefit of this process should be made explicit to all stakeholders
- Try to embed evaluation and assessment to the extent possible and find authentic ways to assess student learning. Find ways so that teachers are given student assessment materials that address their needs, and that evaluations can be conducted in an enjoyable and playful way. Consider learning progressions and follow-up (such as preparation for future learning) when defining goals and outcomes.
- Collaborate with other informal institutions that have similar goals and face similar problems. Working with others improves your ability to involve the formal sector and provides more options for creating lifelong learning pathways for students.
- *Is there a way to enhance interactivity in your setting by using technology and cutting across platforms?* For example, could a museum visit lead people to a website or a real-world setting in which they could continue to explore what they just learned? Could cell phones, MP3 players, or other devices be used to enhance the experience? Are there other ways to use technology to link experiences at informal science environments? Can you capture visitor experiences and provide opportunities for visitors to reflect on their experiences, either onsite or online?

For Further Reading

Anderson, D., Kisiel, J., and Storksdieck, M. (2006). School field trip visits: understanding the teacher's world through the lens of three international studies. *Curator—The Museum Journal, 49*(3), 365-386.

Baldassari, C. (2008). *LabVenture!: At the Cohen Center for Interactive Learning. Summative evaluation report.* Program Evaluation and Research Group, Lesley University.

DeWitt, J., and Storksdieck, M. (2008). A short review on school field trips: Key findings from the past and implications for the future. *Visitor Studies, 11*(2), 181-197.

Falk, J.H., and Dierking, L.D. (1997). School field trips: Assessing their long-term impact. *Curator, 40,* 211-218.

Griffin, J., and Symington, D. (1997). Moving from task-oriented to learning-oriented strategies on school excursions to museums. *Science Education, 81*(6), 763-779.

Koran, J.J., Koran, M.L., and Ellis, J. (1989). Evaluating the effectiveness of field experiences: 1939-1989. *Scottish Museum News, 4*(2), 7-10.

Kubota, C.A., and Olstad, R.G. (1991). Effects of novelty-reducing preparation on exploratory behavior and cognitive learning in a science museum setting. *Journal of Research in Science Teaching, 28*(3), 225-234.

National Research Council (2009). Science learning in designed settings. Chapter 5 in Committee on Learning Science in Informal Environments, *Learning science in informal environments: People, places, and pursuits.* P. Bell, B. Lewenstein, A.W. Shouse, and M.A. Feder (Eds.). Center for Education, Division of Behavioral Sciences and Social Science and Education. Washington, DC: The National Academies Press.

National Science Board (2007). Science, technology, engineering, and mathematics (STEM) education issues and legislative options. In R. Nata (Ed.), *Progress in education* (Vol. 14, pp. 161-189).

Price, S., and Hein, G.E. (1991). More than a field trip: Science programmes for elementary school groups at museums. *International Journal of Science Education, 13*(5), 505-519.

Storksdieck, M., Robbins, D., and Kreisman, S. (2007). *Results from the Quality Field Trip Study: Assessing the LEAD program in Cleveland, Ohio.* Cleveland: Summit Proceedings; University Circle, Inc.

Yager, R.E., and Falk, J. (Eds.). (2008). *Exemplary science in informal education settings: Standards-based success stories.* Arlington, VA: NSTA Press.

Web Resources

LabVenture!: http://mystery.gmri.org/about/default.aspx

LASER: http://www.nsrconline.org/school_district_resources/faqs.html

MERITO: http://montereybay.noaa.gov/educate/merito/outreach-community.html

National Marine Sanctuaries News & Events: Innovative Education Program Heightens Ocean Awareness: http://sanctuaries.noaa.gov/news/features/0706_merito.html

Pacific Science Center: http://www.pacsci.org/

BOX 9-1
MUSEUM 2.0: THE TREND OF THE FUTURE

Designers of children's science programs strive to encourage viewers to make use of multiple platforms to learn about science. After watching a science television show, they hope that viewers consider visiting a local science center or go online to learn more a topic. As technology grows more sophisticated, other informal science venues, such as museums, are providing incentives for their visitors to take advantage of multiple platforms for learning. They are doing so by adding interactive features to their websites, offering visitors a chance to view collections online, view webcasts of special events, respond to blogs, watch videos on YouTube, and receive quick updates about museum events on Twitter.

Museums are approaching this new world in different ways and at different rates. The director of the Bay Area Discovery Museum, a small children's museum in northern California, has started a blog for her museum and engages frequently with Yelp, a Web 2.0 parenting and recreation site. The Library of Congress has posted some images from its collection on Flickr, and the North Carolina Museum of Life and Sciences is experimenting with how to implement Web 2.0 strategies on a small scale.

Larger institutions also are in different stages of developing a strong online and interactive presence. The Smithsonian Institution currently is figuring out how it can become "Smithsonian 2.0." Plans for this institution-wide initiative include digitizing all of the objects in its vast collection, using Facebook to build interest in the Smithsonian, and encouraging visitors to participate in Smithsonian planning by posting their ideas on one of the institution's blogs. The Smithsonian also hopes to change its culture so that the institution no longer sees itself as an "expert" that educates the public, but as a partner that willingly exchanges information with the public and discusses ideas.

The Exploratorium has evolved from posting blogs and exhibits online to building a virtual world that offers visitors a different kind of science experience. In a new world called Exploratorium in Second Life, guests are invited to develop an avatar (a representation of a person) and explore phenomena in ways that are not possible in real life. For example, as part of an exhibit on a solar eclipse, an avatar can literally crawl inside the eclipse's umbra. Avatars also filled an online amphitheatre to share their thoughts about eclipses with their fellow avatars and an Exploratorium (avatar) staff member. And if a visitor wants to talk directly to someone participating in Second Life, tools ranging from instant messaging to online chats are available as well.

These innovations are still in their formative stages, so at this point, research on their impact on learning is not available. But the Exploratorium, the Smithsonian, and many other institutions plan to continue to build their online presence. As they do, the informal science community will develop a deeper understanding of how cutting across multiple platforms and making use of the newest technologies affect learning.

Online presence, use of cell or smart phones, outreach programs into the community, collaborations between schools and informal settings are all initiatives that recognize learning as personal, ongoing, customized, and not constrained by time or settings. In this way, the learner, rather than the provider or the institution, is becoming the central concern in the design process.

BOX 9-2
Major Research Investments into the connection of Formal and Informal Science Teaching and Learning

The National Science Foundation has invested more than $60 Million in the last seven years into four major initiatives that investigate the connection between formal and informal science learning. In addition, a variety of smaller research and development projects across a range of NSF programs have studied this intersection.

By far the largest of these projects is the "Learning in Informal and Formal Environments (LIFE) Center" which seeks to understand and advance human learning through a simultaneous focus on implicit, informal, and formal learning. The goal of research conducted by LIFE is to produce interdisciplinary theories that can guide the design of effective new learning technologies and environments. The LIFE Center brings together experts from research traditions that have so far tended to work separately from one another: neurobiology and psychology, social and cultural sciences, and science learning technologies. A central premise of the LIFE Center is that successful efforts to understand and propel learning require a simultaneous emphasis on informal and formal learning environments, and on the implicit ways in which people learn. The basic research at the LIFE Center is being conducted through three intersecting and multidisciplinary strands of inquiry. The first strand, Implicit Learning and the Brain, investigates the underlying neural processes and principles associated with implicit forms of cognitive, linguistic, and social learning. The second strand, Informal Learning, conducts studies of STEM learning in informal settings to develop comprehensive and coordinated accounts of the cognitive, social, affective, and cultural dimensions that propel learning and development outside of schools. The third strand, Designs for Formal Learning and Beyond, conducts experimental studies in support of designing high-quality learning environments, including theories and measures of transfer (i.e., the ability to utilize what has been learned in one setting, situation or for one problem to another, related one). The $25 M project unites researchers from a variety of universities and non-profit educational research centers.

In 2006, NSF funded a new initiative entitled Academies for Young Scientists with about $14 M. The NSF AYS program funded 15 new projects across the United States, each designed to engage K-8 students to become or remain excited about STEM disciplines. Each of the individual projects is built on partnerships of formal and informal education providers, private sector partners and Colleges of Education to expose students to innovative out-of-school time (OST) learning experiences that demonstrate effective synergies with in-school curricula, and take full advantage of the special attributes of each educational setting in synergistic ways. While projects funded through NSF AYS differ considerably in their individual approaches and desired outcomes (beyond creating excitement and motivation in the youth participants), NSF also provided support for a Learning and Youth Research and Evaluation Center (LYREC) that compares the relative effectiveness of the various implementation models in urban, rural and suburban settings representing diverse student populations. The NSF AYS portfolio of projects, taken as a whole, is designed to inform NSF and the broader educational community of what works and what does not, for whom, and in what setting. LYREC is a collaboration of the Exploratorium, Harvard University, Kings College London, SRI International and UC Santa Cruz. LYREC provides technical assistance to NSF AYS projects, collects and synthesizes their impact data, and oversees dissemination of progress and results. This center builds on the Center for Informal Learning in Schools (CILS) that has developed a theoretical approach that takes into account the particular strengths and affordances of both Out of School Teaching (OST) and school environments.

In 2002, the Center for Informal Learning and Schools (CILS) was funded with almost $12 Million in funding by the National Science Foundation to create a program of research, scholarship, and leadership in the arena of informal learning and the relationship of informal science institutions and schools. CILS is a collaborative effort between the Exploratorium in San Francisco, the University of California at Santa Cruz, and King's College in London (UK). CILS focused its efforts on developing a new crop of scholars and to disseminate its research broadly into the community. Trough dozens of doctoral students and post-doctoral fellows, CILS expanded the area of scholarship in the intersection of formal and informal science education and offered professional development for existing informal science professionals to better enable them to support teachers, students and the general public. Part of CILS, the "Bay Area Institute" has served as a central focus for all CILS activities and helped in disseminating its work to current and future leaders in the field.

CILS focused on making K-12 science education more compelling and accessible to a diverse student population, including students who come from families with little formal experience with K-12 schools and science learning. CILS did this through studying science learning in out-of-school settings, including informal science institutions, and building programmatic bridges between out-of-school and school science learning with the ultimate goal of strengthening alliances between informal learning institutions and schools and broadening conceptions of learning and science learning.

A different perspective on researching the intersection of formal and informal science learning and teaching was taken by the St. Louis Center for Inquiry in Science Teaching and Learning (CISTL), a project supported by more than $10 Million of NSF funding. CISTL combines a focus on research into science teaching and learning with a focus on professional development and support needed to bring inquiry-based teaching and learning into K-12 science education in both formal and informal settings. The project brought together three informal science institutions, two universities, five school districts, one community college system and the Association of Science-Technology Centers (ASTC). CISTL's research agenda focused on the effect of varying types of collaboration and the interfaces among the collaborators (education and scientific; formal and informal) on professional development of new and experienced educators. Part of the project was the development of a diagnostic tool for assessing strengths and weaknesses in science and inquiry backgrounds for teachers and other science educators. Like LIFE and CILS, CISTL aims at synergy between research and practice through research based in practice, practice based on research and the translation of results into practical suggestions for educators.

Aside from these large research to practice initiatives, NSF (and other federal and private funders) have supported a wide variety of projects that link teaching and learning in formal and informal environments. One particular example that might have implications for practice is the almost $1 M project Informal Learning and Science in Afterschool: A Research and Dissemination Project (ILSA). The ILSA research project investigates the nature of informal science in afterschool programs around the country. The three-year study consists of surveys of 1,000 programs, in-depth interviews with a subset of 50, and case studies at eight sites. The study seeks to document the nature of student participation and learning in science activities in "typical" (non-science-specific) afterschool programs, and the infrastructure required to support these programs. "Infrastructure" includes curriculum, staff recruitment and support, and program leadership and structures. The study is brings together researchers at Harvard University (McLean Hospital), the Exploratorium, the Lawrence Hall of Science, and Reginald Clark and Associates. Most importantly, ILSA is part of the Program in Education, Afterschool & Resiliency (PEAR) which is dedicated to making meaningful theoretical and practical contributions to youth development, school reform and prevention. PEAR was founded in 1999 as a collaboration between Harvard Medical School/McLean Hospital and the Harvard Graduate School of Education with a number of strong community partners. The program was established in response to the growing recognition that high-quality afterschool programs hold the promise of building resiliency and preventing high-risk behavior in youth, as well as contributing to school success. PEAR takes a developmental approach to the study of new models of effective afterschool programming, and incorporates educational, health, public policy, and psychological perspectives. PEAR features on its website an assessment tool to measure performance of informal and out-of-school science, technology, engineering and math programs (http://atis.pearweb.org/) that features a broad range of proven methodologies and instruments.

Each of the five featured initiatives (LIFE, CILS, AYS, CISTL and PEAR) publishes their findings through peer-reviewed research articles, conference presentations, symposia, white papers etc., some of which is easily accessible through their informative websites. Yet, like many initiatives of these kinds, transfer of knowledge from original research to practice remains challenging. However, readers are encouraged to look for more information and to connect to the growing network of scholars and scholarly practitioners that emerge from these important investments into the intersection of formal and informal teaching and learning.

Appendixes

APPENDIX A
Biographical Sketches of Oversight Group and Authors

Sue Allen is director of Visitor Research and Evaluation at the Exploratorium in San Francisco. She oversees all aspects of visitor studies, educational research, and evaluation on projects involving the museum's public space. She was the in-house evaluation coordinator on the California Framework Project, which explored the roles that a science museum can play in assisting science education reform in the schools. She and her colleagues studied visitors' learning in the museum's public space and worked collaboratively with practitioners in the design of their research and evaluation agendas. Her current research interests include methods for assessing learning, exhibit design, personal meaning-making, and scientific inquiry. She has lectured in the Department of Museums Studies at John F. Kennedy University. She teaches a graduate-level, action-oriented course on thinking and learning in science to Ph.D. and master's/science credential students in the School of Education at the University of California, Berkeley. She has contributed numerous articles and book chapters to the informal science field. She is a member of many professional associations, including the Visitor Studies Association, the Museum Education Roundtable, and Cultural Connections. She has a Ph.D. in science education from the science and mathematics education program at the University of California, Berkeley.

Marilyn Fenichel *(Author)* is an education writer and editor with writing, editing, and project management expertise. She has worked full-time for J.B. Lippincott, the National Geographic Society, and the National Science Resources Center. She has worked with corporations and nonprofits, textbook companies and multimedia venues, including the National Geographic Society and Discovery Communications. She has written newsletters, film scripts, position papers, website content, annual reports, and catalog copy. She also developed educational products, including lesson plans, activity sheets, facilitators' guides, and teachers' guides. One of her most challenging projects was writing *Science for All Children*, a book on science education reform for the National Science Resources Center, an agency associated with the Smithsonian Institution in Washington, DC. She is a member of the National Association of Science Writers and has participated in conferences sponsored by the National Science Teachers Association and Washington Independent Writers. She graduated cum laude from Bryn Mawr College with an A.B. in English literature.

Myles Gordon works with museums, science centers, and other informal learning organizations around the country, including the California Academy of Sciences, the Smithsonian National Museum of Natural History, the Louisiana State Museum, and Sesame Workshop on strategic planning, educational programming, and outreach. Current projects include strategic plans for education for museums, a human origins education initiative, and a major website on ocean science, education, and conservation. In 1995-2006, he served as vice president for education at the American Museum of Natural History, where he was responsible for programming and product development for children and families, youth, and adults at the museum and in the schools, the community, and nationwide outreach; he also handled production for the Hayden

Planetarium. Prior to joining the museum, he was senior vice president of the Education Development Center, an education research and development group, and director of the Center for Learning and Technology, responsible for work focused on science and math education and technology.

Leslie Herrenkohl is associate professor in the learning sciences and human development and cognition programs in the College of Education at the University of Washington. She also teaches in the Elementary Master's in Teaching Program. She studies children's developing epistemologies of science in formal and informal settings. She also works with practitioners to apply developmental theory to support the design of learning environments. She has completed projects supported by the National Science Foundation, the Spencer Foundation, and the James S. McDonnell Foundation. She has a Ph.D. in psychology from Clark University.

Gil Noam is director of the Program in Education, Afterschool and Resiliency (PEAR) and an associate professor at Harvard Medical School and McLean Hospital. Trained as a clinical and developmental psychologist and psychoanalyst in both Europe and the United States, he has a strong interest in supporting resilience in youth, especially in educational settings. He served as the director of the Risk and Prevention Program and is the founder of the RALLY Prevention Program, a Boston-based intervention that bridges social and academic support in school, after-school, and community settings. He has also followed a large group of high-risk children into adulthood in a longitudinal study that explores clinical, educational, and occupational outcomes. He has published numerous papers, articles, and books in the areas of child and adolescent development as well as risk and resiliency in clinical, school, and after-school settings. He is the editor-in-chief of the journal *New Directions in Youth Development: Theory, Practice and Research*, which has a strong focus on out-of-school time.

Natalie Rusk is an educational researcher and developer in the Media Laboratory at the Massachusetts Institute of Technology, working on the development of Scratch, a new programming language designed for use in community after-school centers. She also works on the design team for Crickets, small programmable devices children can use to create artistic interactive inventions. She specializes in developing technology-based programs and materials that enable young people to create projects based on their interests. She served as project director of the PIE Network, collaborating with the MIT Media Lab and six museums to create a new generation of hands-on science activities that integrate art, crafts, and computer programming. She worked for more than 10 years for the Science Museum of Minnesota, establishing the Learning Technologies Center and guiding the development of the Thinking Fountain and other informal science education websites. In 1993, she cofounded the Computer Clubhouse, a model after-school learning program that engages young people in creating projects with the support of adult mentors. She is interested in how creative applications of technology can help parents and educators support children's positive emotional development. She is a graduate student studying Developmental Technologies in the Eliot-Pearson Department at Tufts University.

Bonnie Sachatello-Sawyer is the founder and director of Hopa Mountain, Inc., a nonprofit organization based in Bozeman, Montana, that is dedicated to supporting rural and tribal community leaders—adults and youth—in their efforts to improve education, ecological health, and economic development. Previously she served as the director of Native Waters at Montana State University and the division head for public programs at the Museum of the Rockies. She is the author or coauthor of four education books, including *Adult Museum Programs: Designing Meaningful Experiences.* She has worked with nonprofit organizations and museums for 18 years as a consultant, facilitator, and guest presenter. She has a Ph.D. in education from Montana State University, an M.A. in political science from the University of Richmond, and a B.A. in political science from Vanderbilt University.

Dennis Schatz is senior vice president for strategic programs at Pacific Science Center in Seattle. A research solar astronomer prior to his career in science education, he worked at the Lawrence Hall of Science at the University of California, Berkeley, prior to moving to Seattle in 1977. He provides leadership to Pacific Science Center's science education programs, which includes a broad range of programs serving teachers, students, community-based organizations, and families across Washington state. He codirects Washington State Leadership and Assistance for Science Education Reform (LASER), a program to implement a quality K-12 science program in all 296 school districts in the state. He has served as principal investigator for a number of National Science Foundation (NSF) projects, including the science center's innovative community leadership project, which develops science advocates in community-based organizations, and the nationally touring exhibit, Aliens: Worlds of Possibilities, which explores the nature of the solar system and the search for extraterrestrial life in the galaxy. He is active in the Association of Science-Technology Centers, being a past member of the program committee, the professional development committee, and past chair of the education committee; he now serves as chair of its Leading Edge Awards Selection Committee. He is also active in the National Science Teachers Association, having been program or general chair for three of its conventions. He is past president of the Astronomical Society of the Pacific.

Dr. Heidi Schweingruber is the Deputy Director of the Board on Science Education at the National Research Council. She has been involved in all of the major projects of the board since it was formed in 2004 and has presented widely on the board's work. She served as study director for a congressionally mandated review of NASA's pre-college education programs and co-directed the study that produced the 2007 report *Taking Science to School: Learning and Teaching Science in Grades K-8.* She was a primary author on the practitioner's version of this report titled, *Ready, Set, Science! Putting Research to Work in K-8 Science Classrooms (2008)* which won a 2008 distinguished achievement award from the Association of Educational Publishers for resources in professional development. She also served as a research associate on *America's Lab Report: Investigations in High School Science (2005).* Prior to joining the National Research Council, Dr. Schweingruber worked as a senior research associate at the Institute of Education Sciences in the U.S. Department of Education. In that role, she served as a program officer for the preschool curriculum evaluation program and for a grant program in mathematics education. Previously, she was the director of research for the Rice University School Mathematics Project, an outreach program in K-12 mathematics education which serves schools and districts in the greater Houston area and taught in the psychology and education departments at Rice University Dr. Schweingruber holds a Ph.D. in psychology (developmental) and anthropology, and a certificate in culture and cognition from the University of Michigan.

Appendix B
Notes

Chapter 1

[1] Jackson, P.W. (1968). Life in classrooms. In A. Pollard and J. Bourne (Eds.), *Teaching and learning in the primary school.* New York: RoutledgeFalmer.

Sosniak, L. (2001). The 9% challenge: Education in school and society. *Teachers College Record, 103,* 15.

[2] Center on Education Policy. (2008). *Instructional time in elementary schools: A closer look at changes for specific subjects.* Washington, DC: Author.

[3] Lemke, J.L. (1992). *The missing context in science education: Science.* Paper presented at the Annual Meeting of the American Education Research Association Conference, San Francisco.

Newton, P., Driver, R., and Osborne, J. (1999). The place of argumentation in the pedagogy of school science. *International Journal of Science Education, 21*(5), 553-576.

National Research Council. (2007). *Taking science to school: Learning and teaching science in grades K-8.* Committee on Science Learning, Kindergarten Through Eighth Grade. R.A. Duschl, H.A. Schweingruber, and A.W. Shouse (Eds.). Washington, DC: The National Academies Press.

Rudolph, J.L. (2002). Scientists in the classroom: The cold war reconstruction of American science education. New York: Palgrave.

[4] Falk, J.H., and Dierking, L.D. (2000). *Learning from museums: Visitor experiences and the making of meaning.* Walnut Creek, CA: AltaMira Press.

Griffin, J. (1998). *School-museum integrated learning experiences in science: A learning journey.* Unpublished doctoral dissertation, University of Technology, Sydney.

[5] National Research Council. (2009). *Learning Science in Informal Environments: People, Places, and Pursuits.* Committee on Learning Science in Informal Environments. Philip Bell, Bruce Lewenstein, Andrew W. Shouse, and Michael A. Feder, editors. Board on Science Education, Center for Education, Division of Behavioral and Social Sciences and Education. Washington, DC: The National Academies Press.

[6] National Research Council. (1999). *How people learn: Brain, mind, experience, and school.* Committee on Developments in the Science of Learning. J.D. Bransford, A.L. Brown, and R.R. Cocking (Eds.). Washington, DC: National Academy Press.

[7] National Research Council. (2000). *How people learn: Brain, mind, experience, and school* (expanded ed.). Committee on Developments in the Science of Learning. J.D. Bransford, A.L. Brown, and R.R. Cocking (Eds.). Washington, DC: National Academy Press.

Chapter 2

[1] Aikenhead, G. (1996). Science education: Border crossing into the subculture of science. *Studies in Science Education, 27*, 1-52.

[2] Storksdieck, M., and Falk, J.H. (2004). Evaluating public understanding of research projects and initiatives. In D. Chittendan, G. Farmelo, and B.V. Lewenstein (Eds.), *Creating connections—Museums and the public understanding of research* (pp. 87-108). Walnut Creek, CA: AltaMira Press.

[3] National Research Council. (2007). *Taking science to school: Learning and teaching science in grades K-8.* Committee on Science Learning, Kindergarten Through Eighth Grade. R.A. Duschl, H.A. Schweingruber, and A.W. Shouse (Eds.). Washington, DC: The National Academies Press.

[4] Jolly, E., Campbell, P., and Perlman, L. (2004). *Engagement, capacity, continuity: A trilogy for student success.* St. Paul: GE Foundation and Science Museum of Minnesota.

Tai, R.H., Liu, C.Q., Maltese, A.V., and Fan, X. (2006) *Planning early for careers in science.* Science *312,* 1143-1144.

[5] National Research Council. (2000). *How people learn: Brain, mind, experience, and school* (expanded ed.). Committee on Developments in the Science of Learning. J.D. Bransford, A.L. Brown, and R.R. Cocking (Eds.). Washington, DC: National Academy Press.

[6] Osborne, J., Collins, S., Ratcliffe, M., Millar, R., and Duschl, R. (2003). What "ideas about science" should be taught in school science? A Delphi study of the expert community. *Journal of Research in Science Teaching, 40*(7), 692-720.

[7] American Association for the Advancement of Science. (1993). *Benchmarks for science literacy.* New York: Oxford University Press.

Chapter 3

[1] Crowley, K., and Jacobs, M. (2002). Islands of expertise and the development of family scientific literacy. In G. Leinhardt, K. Crowley, and K. Knutson (Eds.), *Learning conversations in museums*. Mahwah, NJ: Lawrence Erlbaum.

Reeve, S., and Bell, P. (in press). Children's self-documentation and understanding of the concepts "healthy" and "unhealthy." *International Journal of Science Education.*

[2] National Research Council. (1999). *How people learn: Brain, mind, experience, and school*. Committee on Developments in the Science of Learning. J.D. Bransford, A.L. Brown, and R.R. Cocking (Eds.). Washington, DC: National Academy Press.

[3] Allen, S. (1997). Using scientific inquiry activities in exhibit explanations. *Science Education, 81*(6), 715-734.

Borun, M., and Miller, M. (1980). *What's in a name? A study of the effectiveness of explanatory labels in a science museum.* Philadelphia: Franklin Institute Science Museum.

Peart, B. (1984). Impact of exhibit type on knowledge gain, attitudes, and behavior. *Curator, 27*(3), 220-227.

Chapter 4

[1] Blum-Kulka, S. (1997). *Dinner talk: Cultural patterns of sociability and socialization in family discourse*. Mahwah, NJ: Lawrence Erlbaum Associates.

Callanan, M.A., Shrager, J., and Moore, J. (1995). Parent-child collaborative explanations: Methods of identification and analysis. *Journal of the Learning Sciences, 4*, 105-129.

[2] Blum-Kulka, S. (2002). Do you believe that Lot's wife is blocking the road (to Jericho)? Co-constructing theories about the world with adults. In S. Blum-Kulka and C.E. Snow (Eds.), *Talking to adults: The contribution of multiparty discourse to language acquisition* (pp. 85-116). Mahwah, NJ: Lawrence Erlbaum Associates.

Tenenbaum, H.R., and Callanan, M.A. (2008). Parents' science talk to their children in Mexican-descent families residing in the United States. *International Journal of Behavioral Development, 32*(1), 1-12.

[3] Callanan, M.A., and Oakes, L. (1992). Preschoolers' questions and parents' explanations: Causal thinking in everyday activity. *Cognitive Development, 7*, 213-233.

Chouinard, M.M. (2007). Children's questions: A mechanism for cognitive development. *Monographs of the Society for Research in Child Development, 72*(1), 1-121.

[4] National Research Council. (2007). *Taking science to school: Learning and teaching science in grades K 8.* Committee on Science Learning, Kindergarten Through Eighth Grade. R.A. Duschl, H.A. Schweingruber, and A.W. Shouse (Eds.). Washington, DC: The National Academies Press

[5] Moll, L., Amanti, C., Neff, D., and Gonzalez, N. (2005). Funding of knowledge for teaching: Using a qualitative approach to connect homes and classrooms. In L. Moll, C. Amanti, and N. Gonzalez (Eds.), *Funds of knowledge: Theorizing practices in households, communities, and classrooms* (pp. 71-88). London: Routledge.

Palmquist, S., and Crowley, K. (2007). From teachers to testers: How parents talk to novice and expert children. *Science Education, 91*(5), 783-804.

[6] Fienberg, J., and Leinhardt, G. (2002). Looking through the glass: Reflections of identity in conversations at a history museum. In G. Leinhardt, K. Crowley, and K. Knutson (Eds.), *Learning conversations in museums* (pp. 167-211). Mahwah, NJ: Lawrence Erlbaum Associates.

[7] Koran, J.J., Koran, M.L., and Foster, J.S. (1988). Using modeling to direct attention. *Curator, 31*(1), 36-42.

[8] Saltz, C., Crocker, N., and Banks, B. (2004). *Evaluation of service at the Salado for fall 2004.* Tempe: Arizona State University International Institute for Sustainability. Available: http://caplter.asu.edu/explorers/riosalado/pdf/fall04_report.pdf [accessed October 2008].

[9] Cazden, C.B. (2001). Classroom discourse: The language of teaching and learning (2nd ed.). Westport, CT: Heinemann.

National Research Council. (2007). *Taking science to school: Learning and teaching science in grades K 8.* Committee on Science Learning, Kindergarten Through Eighth Grade. R.A. Duschl, H.A. Schweingruber, and A.W. Shouse (Eds.). Washington, DC: The National Academies Press

[10] Gee, J.P. (1994). First language acquisition as a guide for theories of learning and pedagogy. *Linguistics and Education, 6*(4), 331-354.

Lemke, J.L. (1990). *Talking science: Language, learning and values*. Norwood, NJ: Ablex.

National Research Council. (2007). *Taking science to school: Learning and teaching science in grades K 8.* Committee on Science Learning, Kindergarten Through Eighth

Grade. R.A. Duschl, H.A. Schweingruber, and A.W. Shouse (Eds.). Washington, DC: The National Academies Press

[11] Callanan, M.A., and Oakes, L. (1992). Preschoolers' questions and parents' explanations: Causal thinking in everyday activity. *Cognitive Development*, *7*, 213-233.

Chouinard, M.M. (2007). Children's questions: A mechanism for cognitive development. *Monographs of the Society for Research in Child Development, 72*(1), 1-121.

[12] National Research Council. (2007). *Taking science to school: Learning and teaching science in grades K 8.* Committee on Science Learning, Kindergarten Through Eighth Grade. R.A. Duschl, H.A. Schweingruber, and A.W. Shouse (Eds.). Washington, DC: The National Academies Press

Chapter 5

[1] Falk, J.H., and Storksdieck, M. (2005). Using the contextual model of learning to understand visitor learning from a science center exhibition. *Science Education*, *89*, 744-778.

[2] Nasir, N.S., Rosebery, A.S., Warren B., and Lee, C.D. (2006). Learning as a cultural process: Achieving equity through diversity. In R.K. Sawyer (Ed.), *The Cambridge handbook of the learning sciences* (pp. 489-504). New York: Cambridge University Press.

[3] National Research Council. (2000). *How people learn: Brain, mind, experience, and school* (expanded ed.). Committee on Developments in the Science of Learning. J.D. Bransford, A.L. Brown, and R.R. Cocking (Eds.). Washington, DC: National Academy Press.

[4] Falk, J., and Dierking, L.D. (2000). *Learning from museums.* Walnut Creek, CA: AltaMira Press.

[5] Hayward, J. (1997). Conservation study, phase 2: An analysis of visitors' perceptions about conservation at the Monterey Bay Aquarium. Northampton, MA: People, Places, and Design Research.

Yalowitz, S.S. (2004). Evaluating visitor conservation research at the Monterey Bay Aquarium. *Curator*, *47*(3), 283-298.

[6] Myers, G., Saunders, C.D., and Birjulin, A.A. (2004). Emotional dimensions of watching zoo animals: An experience sampling study building on insights from psychology. *Curator*, *47*, 299-321.

[7] Csikszentmihalyi, M., Rathunde, K., and Whalen, S. (1993). *Talented teenagers: The roots of success and failure.* New York: Cambridge University Press.

Lipstein, R., and Renninger, K.A. (2006). "Putting things into words": The development of 12-15-year-old students' interest for writing. In P. Boscolo and S. Hidi (Eds.), *Motivation and writing: Research and school practice* (pp. 113- 140). New York: Kluwer Academic/Plenum.

Renninger, K.A., and Hidi, S. (2002). Interest and achievement: Developmental issues raised by a case study. In A. Wigfield and J. Eccles (Eds.), *Development of achievement motivation* (pp. 173-195). New York: Academic Press.

[8] Lipstein, R., and Renninger, K.A. (2006). "Putting things into words": The development of 12-15-year-old students' interest for writing. In P. Boscolo and S. Hidi (Eds.), *Motivation and writing: Research and school practice* (pp. 113- 140). New York: Kluwer Academic/Plenum.

Renninger, K.A., and Hidi, S. (2002). Interest and achievement: Developmental issues raised by a case study. In A. Wigfield and J. Eccles (Eds.), *Development of achievement motivation* (pp. 173-195). New York: Academic Press.

Renninger, K.A., Sansone, C., and Smith, J.L. (2004). Love of learning. In C. Peterson and M.E.P. Seligman (Eds.), *Character strengths and virtues: A handbook and classification* (pp. 161-179). New York: Oxford University Press.

[9] Falk, J.H., Reinhard, E.M., Vernon, C.L., Bronnenkant, K., Deans, N.L., and Heimlich, J.E. (2007). *Why zoos and aquariums matter: Assessing the impact of a visit.* Silver Spring, MD: Association of Zoos and Aquariums.

[10] Holland, D., Lachicotte, W., Skinner, D., and Cain, C. (1998). *Identity and agency in cultural worlds.* Cambridge, MA: Harvard University Press.

Hull, G.A., and Greeno, J.G. (2006). Identity and agency in nonschool and school worlds. In Z. Bekerman, N. Burbules, and D.S. Keller (Eds.), *Learning in places: The informal education reader* (pp. 77-97). New York: Peter Lang.

[11] Beane, D.B., and Pope, M.S. (2002). Leveling the playing field through object-based service learning. In S.G. Paris (Ed.), *Perspectives on object-centered learning in museums* (pp. 325-349). Mahwah, NJ: Lawrence Erlbaum Associates.

McCreedy, D. (2005). Engaging adults as advocates. *Curator, 48*(2), 158-176.

[12] Anderson, D. (2003). Visitors' long-term memories of world expositions. *Curator, 46*(4), 401-420.

Falk, J.H. (2006). The impact of visit motivation on learning: Using identity as a construct to understand the visitor experience. *Curator, 49*(2), 151-166.

Leinhardt, G., and Knutson, K. (2004). *Listening in on museum conversations.* Walnut Creek, CA: AltaMira Press.

National Research Council. (2007). *Taking science to school: Learning and teaching science in grades K-8.* Committee on Science Learning, Kindergarten Through Eighth Grade. R.A. Duschl, H.A. Schweingruber, and A.W. Shouse (Eds.). Washington, DC: The National Academies Press.

[13] Roth, E.J., and Li, E. (2005, April). *Mapping the boundaries of science identity in ISME's first year.* A paper presented at the annual meeting of the American Educational Research Association, Montreal.

Weinburgh, M.H., and Steele, D. (2000). The modified attitude toward science inventory: Developing an instrument to be used with fifth grade urban students. *Journal of Women and Minorities in Science and Engineering, 6*(1), 87-94.

[14] Brown, B., Reveles, J., and Kelly, G. (2004). Scientific literacy and discursive identity: A theoretical framework for understanding science learning. *Science Education, 89*(5), 779-802.

Hull, G.A., and Greeno, J.G. (2006). Identity and agency in nonschool and school worlds. In Z. Bekerman, N. Burbules, and D.S. Keller (Eds.), *Learning in places: The informal education reader* (pp. 77-97). New York: Peter Lang.

Holland, D., and Lave, J. (Eds.) (2001). *History in person: Enduring struggles, contentious practice, intimate identities.* Albuquerque, NM: School of American Research Press.

Jacoby, S., and Gonzales, P. (1991). The constitution of expert-novice in scientific discourse. *Issues in Applied Linguistics, 2*(2), 149-181.

Rounds, J. (2006). Doing identity work in museums. *Curator, 49*(2), 133-150.

[15] Hull, G.A., and Greeno, J.G. (2006). Identity and agency in nonschool and school worlds. In Z. Bekerman, N. Burbules, and D.S. Keller (Eds.), *Learning in places: The informal education reader* (pp. 77-97). New York: Peter Lang.

Jacoby, S., and Gonzales, P. (1991). The constitution of expert-novice in scientific discourse. *Issues in Applied Linguistics, 2*(2), 149-181.

Brown, B., Reveles, J., and Kelly, G. (2004). Scientific literacy and discursive identity: A theoretical framework for understanding science learning. *Science Education, 89*(5), 779-802.

[16] Ellenbogen, K.M. (2003). From dioramas to the dinner table: An ethnographic case study of the role of science museums in family life. *Dissertation Abstracts International*, *64*(3), 846-847.

Chapter 6

[1] Schwartz, D.L., Bransford, J.D., and Sears, D. (2005). Efficiency and innovation in transfer. In J.P. Mestre (Ed.), *Transfer of learning from a modern multidisciplinary perspective* (pp. 1-51). Greenwich, CT: Information Age.

[2] National Research Council. (2002). *Scientific research in education.* Committee on Scientific Principles for Education Research. R.J. Shavelson and L. Towne (Eds.). Washington, DC: National Academy Press.

Chapter 7

[1] Heath, S., (2007). *Diverse learning and learner diversity in "informal" science learning environments.* Background paper for the Committee on Science Education for Learning Science in Informal Environments. Available: http://www7.nationalacademies.org/bose/Learning_Science_in_Informal_Environments_ Commissioned_Papers.html [accessed November 2008].

[2] Allen, G., and Seumptewa, O. (1993). The need for strengthening Native American science and mathematics education. In S. Carey (Ed.), *Science for all cultures: A collection of articles from NSTA's journals* (pp. 38-43). Arlington, VA: National Science Teachers Association.

Banks, J.A. (2007). *Educating citizens in multicultural society* (2nd ed.). New York: Teachers College Press.

Cajete, G. (1993). *Look top the mountain: An ecology of Indian education.* Skyland, NC: Kivaki Press.

MacIvor, M. (1995). Redefining science education for aboriginal students. In M. Battiste and J. Barman (Eds.), *First nations education in Canada: The circle unfolds* (pp. 73-98). Vancouver: University of British Columbia Press.

Malcom, S.M., and Matyas, M.L. (Eds.) (1991). *Investing in human potential: Science and engineering at the crossroads.* Washington, DC: American Association for the Advancement of Science.

Snively, G. (1995). Bridging traditional science and western science in the multicultural classroom. In G. Snively and A. MacKinnon (Eds.), *Thinking globally about mathematics and science education* (pp. 53-75). Vancouver: University of British Columbia, Centre for the Study of Curriculum and Instruction.

[3] Barton, A.C. (2008). Creating hybrid spaces for engaging school science among urban middle school girls. *American Educational Research Journal, 45*(1), 68-103.

[4] Garibay, 2009; Garcia-Luis, 2007; Monaco and Strasser, 2006

[5] Garibay, C. (2004). *Animal secrets bilingual labels formative evaluation.* Unpublished manuscript, Garibay Group, Chicago.

Garibay, C., and Gilmartin, J. (2003). *Chocolate summative evaluation at the Field Museum.* Unpublished manuscript, Garibay Group, Chicago.

[6] Garibay, C. (2004). *Animal secrets bilingual labels formative evaluation.* Unpublished manuscript, Garibay Group, Chicago.

Garibay, C., and Gilmartin, J. (2003). *Chocolate summative evaluation at the Field Museum.* Unpublished manuscript, Garibay Group, Chicago.

[7] Ash, D. (2004). Reflective scientific sense-making dialogue in two languages: The science in the dialogue and the dialogue in the science. *Science Education, 88*, 855-884.

Wheaton, M., and Ash, D. (2008). Exploring middle school girls' ideas about science at a bilingual marine science camp. *Journal of Museum Education, 33*(2), 131-143.

[8] Aikenhead, G. (2001). *Cross-cultural science teaching: Praxis.* A paper presented at the annual meeting of the National Association for Research in Science Teaching, St. Louis, March 26-28.

Chapter 8

[1] Falk, J.H., and Dierking, L.D. (2002). Lessons without limit: How free choice learning is transforming education. Walnut Creek, CA: AltaMira Press.

[2] Baillargeon, R. (2004). How do infants learn about the physical world? *Current Directions in Psychological Science, 3*, 133-140.

Cohen, L.B., and Cashon, C.H. (2006). Infant cognition. In W. Damon and R.M. Lerner (Series Eds.) and D. Kuhn and R.S. Siegler (Vol. Eds.), *Handbook of child psychology: Cognition, perception, and language* (vol. 2, 6th ed., pp. 214-251). New York: Wiley.

[3] Krist, H., Fieberg, E.L., and Wilkening, F. (1993). Intuitive physics in action and judgment: The development of knowledge about projectile motion. *Journal of Experimental Psychology: Learning, Memory, and Cognition, 19*(4), 952.

[4] Chi, M.T.H., and Koeske, R.D. (1983). Network representation of a child's dinosaur knowledge. *Developmental Psychology*, *19*(1), 29-39.

Crowley, K., and Jacobs, M. (2002). Islands of expertise and the development of family scientific literacy. In G. Leinhardt, K. Crowley, and K. Knutson (Eds.), *Learning conversations in museums*. Mahwah, NJ: Lawrence Erlbaum.

[5] National Research Council. (2007). *Taking science to school: Learning and teaching science in grades K-8*. Committee on Science Learning, Kindergarten Through Eighth Grade. R.A. Duschl, H.A. Schweingruber, and A.W. Shouse (Eds.). Washington, DC: The National Academies Press.

[6] Korpan, C.A., Bisanz, G.L., Bisanz, J., and Lynch, M.A. (1998). *Charts: A tool for surveying young children's opportunities to learn about science outside of school.* Ottawa: Canadian Social Science and Humanities Research Council.

[7] Farenga, S.J., and Joyce, B.A. (1997). Beyond the classroom: Gender differences in science experiences. *Education, 117*, 563-568.

[8] Falk, J.H., and Dierking, L.D. (2002). *Lessons without limit: How free choice learning is transforming education*. Walnut Creek, CA: AltaMira Press.

[9] Falk, J.H., and Dierking, L.D. (2002). *Lessons without limit: How free choice learning is transforming education*. Walnut Creek, CA: AltaMira Press.

[10] Zimmer-Gembeck, M.J., and Collins, W.A. (2003). Autonomy development during adolescence. In G.R. Adams and M.D. Berzonsky (Eds.), *Blackwell handbook of adolescence* (pp. 175-204). Malden, MA: Blackwell.

[11] Csikszentmihalyi, M., and Larson, R. (1984). *Being adolescent.* New York: Basic Books.

[12] Eccles, J.S., Lord, S., and Buchanan, C.M. (1996). School transitions in early adolescence: What are we doing to our young people? In J. Graber, J. Brooks-Gunn, and A. Petersen (Eds.), *Transitions through adolescence: Interpersonal domains and context* (pp. 251-284). Mahwah, NJ: Lawrence Erlbaum Associates.

[13] Greenberger, E. and Steinberg, L. (1986). *When teenagers work: The psychological and social costs of adolescent employment.* New York: Basic Books.

Mahoney, J.L., Larson, R.W., and Eccles, J.S. (2005). Organized activities as development contexts for children and adolescents. In J.L. Mahoney, R.W. Larson, and J.S. Eccles (Eds.), *Organized activities as contexts of development: Extracurricular activities, after-school and community programs* (pp. 3-22). Mahwah, NJ: Erlbaum.

[14] Barron, B. (2006). Interest and self-sustained learning as catalysts of development: A learning ecology perspective. *Human Development, 49*(4), 153-224.

[15] Flynn, K.E., Smith, M.A., and Freese, J. (2006). When do older adults turn to the Internet for health information? Findings from the Wisconsin longitudinal study. *Journal of General Internal Medicine, 21*(12), 1295-1301.

[16] Kelly, L., Savage, G., Landman, P., and Tonkin, S. (2002). *Energised, engaged, everywhere: Older Australians and museums.* Canberra: National Museum of Australia.

[17] American Transportation Association (2007). *Transitions to transportation options: How they affect older adults.* http://www.apta.com/research/info/online/transitions.cfm#research.

[18] Kelly, L., Savage, G., Landman, P., and Tonkin, S. (2002). *Energised, engaged, everywhere: Older Australians and museums.* Canberra: National Museum of Australia.

19 (still working on)

[20] Elder, G. H. (1974). *Children of the Great Depression.* Chicago: University of Chicago Press.

Chapter 9

[1] Schauble, L., and Bartlett, K. (1997). Constructing a science gallery for children and families: The role of research in an innovative design process. *Science Education, 81*(6), 781-793.

[2] See various publications by John Falk and Lynn Dierking that discuss interconnected systems for lifelong leanring from a programmatic perspective.

[3] So-called subsequent reinforcing experiences or follow-up that connects individual learning experiences have been researched by a variety of scholars (often as longitudinal research), although evaluation and research still does not fully appreciate the need to understand the connections between individual experiences (how and why chosen, and how linked)

[4] Much of this literature is now accessible online and provided by the informal setting, though there are also comprehensive guides in book form available, such as Kathleen Carroll's 2007 A Guide to Great Field Trips (Zephir Press), which combines practical how-to advice with summaries of scholarly work on field trips.

[5] Kubota, C.A., and Olstad, R.G. (1991). Effects of novelty-reducing preparation on exploratory behavior and cognitive learning in a science museum setting. *Journal of Research in Science Teaching, 28*(3), 225-234.

[6] Price, S., and Hein, G.E. (1991). More than a field trip: Science programmes for elementary school groups at museums. *International Journal of Science Education, 13*(5), 505-519.

[7] Koran, J.J., Koran, M.L., and Ellis, J. (1989). Evaluating the effectiveness of field experiences: 1939-1989. *Scottish Museum News, 4*(2), 7-10.

[8] Griffin, J., and Symington, D. (1997). Moving from task-oriented to learning oriented strategies on school excursions to museums. *Science Education, 81*(6), 763-779.

[9] Burtnyk, K.M., and Combs, D.J. (2005). Parent chaperones as field trip facilitators: A case study. *Visitor Studies Today 8*(1), 13-20. [reference: http://www.informalscience.org/researches/VSA-a0a5z9-a_5730.pdf]

[10] Griffin, J. (1994). Learning to learn in informal science settings. *Research in Science Education, 24*(1), 121-128.

[11] Anderson, D., Kisiel, J., and Storksdieck, M. (2006). Understanding teachers' perspectives on field trips: Discovering common ground in three countries. *Curator, 49*(3), 365-386.

DeWitt, J., and Storksdieck, M. (2008). A short review on school field trips: Key findings from the past and implications for the future. *Visitor Studies, 11*(2), 181-197.

[12] McLaughlin, M. (2000). Community counts: How youth organizations matter for youth development. Washington, DC: Public Education Fund Network.

[13] McLaughlin, M. (2000). Community counts: How youth organizations matter for youth development. Washington, DC: Public Education Fund Network.

[14] Noam, G., Biancarosa, G., and Dechausay, N. (2003). *After-school education: Approaches to an emerging field.* Cambridge, MA: Harvard Education Press.

[15] Anderson, D., Lawson, B., and Mayer-Smith, J. (2006). Investigating the impact of a practicum experience in an aquarium on pre-service teachers. *Teaching Education, 17*(4), 341-353.

[16] RMC Corporation, 2004.

[17] National Research Council. (2006). *Systems for state science assessment.* Committee on Test Design for K-12 Science Achievement. M.R. Wilson and M.W. Bertenthal (Eds.). Board on Testing and Assessment, Center for Education, Division of Behavioral and Social Sciences and Education. Washington, DC: The National Academies Press.

National Research Council. (2007). *Taking science to school: Learning and teaching science in grades K-8.* Committee on Science Learning, Kindergarten Through Eighth Grade. R.A. Duschl, H.A. Schweingruber, and A.W. Shouse (Eds.). Washington, DC: The National Academies Press.